UNIVERSAL F

Dr Sidney Perkowitz is Charles Howard Candler Professor of Physics at Emory University. A renowned expert on the optical properties of matter, he writes for magazines such as *The Sciences* and *New Scientist* and is the author of *Empire of Light: A History of Discovery of Science and Art.* He lives in Atlanta, Georgia.

ALSO BY SIDNEY PERKOWITZ

Empire of Light: A History of Discovery in Science and Art

Sidney Perkowitz

UNIVERSAL FOAM

The story of bubbles from
cappuccino to the cosmos

VINTAGE

Published by Vintage 2001

1 3 5 7 9 10 8 6 4 2

Copyright © Sidney Perkowitz 2000

Sidney Perkowitz has asserted his right under the Copyright, Designs and Patents Act 1988 to be identified as the author of this work

First published in Great Britain by
Vintage 2001

Vintage
Random House, 20 Vauxhall Bridge Road,
London SW1V 2SA

Random House Australia (Pty) Limited
20 Alfred Street, Milsons Point, Sydney,
New South Wales 2061, Australia

Random House New Zealand Limited
18 Poland Road, Glenfield,
Auckland 10, New Zealand

Random House (Pty) Limited
Endulini, 5A Jubilee Road, Parktown 2193,
South Africa

The Random House Group Limited Reg. No. 954009
www.randomhouse.co.uk

A CIP catalogue record for this book
is available from the British Library

ISBN 0 09 928656 4

Papers used by Random House are natural, recyclable products made from wood grown in sustainable forests. The manufacturing processes conform to the environmental regulations of the country of origin.

Printed and bound in Great Britain by
Cox & Wyman Ltd, Reading, Berkshire

TO SANDY AND MIKE,
WITH LOVE AND APPRECIATION

contents

acknowledgments

An author with family, friends, and colleagues who help him write a book is a lucky man. My good fortune begins with my wonderful wife, Sandy, who—as always—was my best reader, helping me write words that make sense.

Universal Foam would not exist without my smart and effective agent, Michael Carlisle, who convinced me that a proposed magazine article could become a book, and then placed the book. At Walker & Company, publisher George Gibson supported *Universal Foam* with more than enthusiasm, even passing on relevant items about foam. Jackie Johnson provided insightful editing that brought out my best writing, backed up by a dedicated production staff.

Emory University provided an excellent environment for writing. Friends and colleagues there drew on their specialties to read drafts, answer questions, and suggest interviewees: Krishan Bajaj, Katherine Benson, David Bright, Ray DuVarney, Candy Lang, and Paul Lennard, with special thanks to Bill Size and Leslie Taylor. Three Emory undergraduates, Matt Aldag, Bon Fleming, and Jessica Whiting, tenaciously carried out research. In my courses "Introductory Astronomy" and "Communicating Science," other students helped by reacting to my demonstrations and ideas about foam. Phil Schewe and Ben Stein at the American Institute of Physics kept me abreast of current foam research. Brian Schwartz of the American Physical Society arranged my public lecture "The Physics of Beer," which sharpened my thinking about foamy food and drink. Michael Sims shared his encyclopedic knowledge, Ann Duerr educated me about biomedicine, and David Schomer of Versace Coffee, Seattle, was helpful. I benefited from the visiting scholars program at the Huntington Library in Pasadena, California, which enabled me to see Robert Hooke's marvelous drawings in a first edition of his *Micrographia*. Of course, any errors in this book are my doing, not that of these people and institutions.

The idea for *Universal Foam* began as I drank espresso and cappuccino. It is only right that much of the book was written at a favorite spot, Atlanta's San Francisco Coffee Shop, where Doug and Tanya Bond and their personable *baristas* kept the coffee foam and the good music coming. Karen, Michael, Randy, and other fellow San Fran habitués provided pleasant diversion between bouts of writing.

I owe a particular "thank you" to the approximately thirty people I interviewed for *Universal Foam*. Scientist and engineer, artist, chef, and businessperson—you gave me generous amounts of time as you patiently answered questions, sharing the knowledge and passion you bring to your pursuits. You have added enormously to this book.

Finally, to my beloved family and dear friends—thank you for gracefully accepting and indulging my foamy obsession.

UNIVERSAL *foam*

INTRODUCTION

I've always liked coffee, but my delight took on a new dimension when I started drinking espresso and cappuccino. My spirits lift every morning as I grind and measure the coffee, put it into the espresso machine, and watch it drip. Then comes the best part: I plunge the steam pipe into milk, making bubbles which cling closely to each other to form the magical froth that turns an espresso into a cappuccino. Sometimes the bubbles are tiny and densely packed, which makes the foam stiff and upstanding. Sometimes they're large and fragile, forming a shaky structure that soon collapses. The variations depend mostly on the type and brand of milk, and the pressure of the steam.

That arrangement of adjoining bubbles is a classic example of foam. I enjoy it while I drink my coffee, yet I could with equal pleasure bring the foam into my physics laboratory for serious study. My specialty is the physics of matter, which I examine with lasers, optical instruments, and computers. And although I have conducted and published research on every kind of matter, from exotic metals to crystalline sapphire, from liquids and gases to complex biological molecules, few of the systems I have studied represent the scientific challenges of foam.

For all its frothiness, foam is a surprisingly intricate formation that has impact on astronomy, biology, chemistry, physics, and mathematics. Foam is not exclusively a solid, liquid, or gas; it is made of bubbles or cells of gas within a liquid or a solid, and it combines characteristics of all three states of matter. David Weitz, a physicist at Harvard University who has studied foam for twenty years, calls foam the "neglected material"; neglected, that is, by scientists. Grasping the complexities of foam and its bubbles requires scientific tools and knowledge that have come along only recently.

Much of what we know about foam we have learned through its use in technology and in commerce. In the food and beverage industries, whole categories of products such as bread, beer, and champagne are built on ephemeral bubbles. In other industries, foam enhances oil-drilling operations and enters into the manufacture of novel forms of metals. The plastics industry depends largely on foam, from polystyrene to polyurethane.

We live in a universe inundated with foam. It may be startling to realize that such an airy substance carries true scientific weight. At the planetary scale, there is the foam of ocean whitecaps that covers millions of square miles and influences the world's climate. Pumice, a type of foamy rock emitted from volcanoes, carries clues to the geologic history of the Earth. At the cosmic scale, the billions of galaxies making up the universe are arranged as if they lay on the surfaces of immense bubbles within a gargantuan foam.

One of the most exotic and intriguing foams is aerogel, a foamlike material made of glass. Slabs of aerogel are now aboard a NASA spacecraft on its way to a rendezvous with a comet called WILD-2. The foamy nature of aerogel gives it a unique ability to capture and hold fast-moving particles emitted by the comet. When this material is returned to Earth in the year 2006, it is expected to give new insight into the origin of the Sun and its planets, and perhaps into the beginnings of life.

Although much is known about foam, it is studded with scientific mysteries. Some represent unexplained properties of foam seen even in everyday use, such as the flowing behavior of whipped cream, a singular combination of solid and liquid characteristics. Others arise in the laboratory, such as the extraordinary effect called sonoluminescence, where a bubble floating in liquid changes sound into light. Still others involve patterns deeply embedded in nature, such as the foamy turmoil in matter and energy that originates in the random nature of quantum events.

Scientists examine these questions with complex theories and an arsenal of experimental tools: laser probes, underwater technology, sophisticated data analysis, and advanced imaging techniques. One of the most powerful of these is the computer. It is used to explore the geometry and dynamic behavior of foam, giving insights difficult to achieve through ordinary experimentation.

Yet you can observe the curious behavior of foam without elaborate equipment. In the nineteenth century, Joseph Antoine Ferdinand Plateau established the geometric principles of foam by studying soap bubbles on wire frames. You can make your own bubble laboratory by adding a bit of liquid soap to some water inside a clear plastic bottle. Shake the bottle, and it fills with a foam of adjoining soap bubbles. Tilt the bottle to see how the foam moves; shine a light through the white mass; notice how the bubbles change with time. Your examination will reveal many of the properties of foam, which can be explained only by combining scientific disciplines—a collective approach I call *foam science,* meaning the study of the behavior and appearance of foam and its patterns by all scientific means.

There is no single center for research based on foam; rather, foam science is carried out around the world, from the Department of Aerospace Engineering at Boston University, to NASA's Jet Propulsion Laboratory in Pasadena, California, to Trinity College, Dublin, to French Polynesia, and a dozen other sites.

Foam's many applications enhance—and even preserve— our lives. Shaving cream keeps beard hair moist, making it easier to cut; foamed plastic protects the fuel tanks on NASA space shuttles (although that once had a completely unintended outcome) and insulates racing car drivers in their cramped cockpits; foamed metal can make better prosthetic devices. Small bubbles injected into the bloodstream allow the medical technique of ultrasound imaging to detect cancer; fire-fighting foam made from soybeans saved countless lives in World War II; and foams under development at the Sandia National Laboratories may serve to muffle explosions set by terrorists.

Applied foam science is also carried out at some unexpected sites. In Reims, France, the heart of champagne country, scientists at the Oenology Laboratory seek the best possible champagne froth. In his kitchen, American biochemist and food expert Harold McGee establishes how to make the best meringues.

One of the charms of foam is its aesthetic appeal. The evanescent beauty of bubbles has put them into works of art, and the association of foam with tossing seas gives it a special place in mythology, which engenders its own art. Aphrodite, Greek goddess of love and fertility, and her Roman counterpart Venus sprang from sea foam, as captured in Botticelli's painting

The Birth of Venus. Botticelli's contemporary Leonardo da Vinci was drawn to foamy turbulence, delighting in drawing huge splashes full of foam and white water. (Being Leonardo, he also made scientific studies that relate to the nature of bubbles.) Artist after artist has shown the power of the sea through foam, as in the celebrated nineteenth-century print *Great Wave off Kanagawa* by Katsushika Hokusai, with its striking tendrils of white foam reaching out from an enormous wave.

Foam opens different doors into science and technology, classical, modern, and future. It also opens doors into human culture, through its relations with food and art, and in a myriad of useful applications. And its very frothiness brings something special: a delight that comes from its fragility, its beauty, its sensual appeal. As *Universal Foam* explores the science of foam and its varied uses, it will heighten your appreciation of how wondrous and widespread foam is.

THE BASICS OF FOAM

BUBBLES AND GEOMETRY

Take a moment to consider the world around you. Look at its varied colors and textures; touch, smell, and even taste it. No doubt you'll soon notice the remarkable diversity of matter that surrounds us. Now group all that you see into solids, liquids, and gases. Even within these categories, there are subgroups. You may be sitting, for instance, on a soft solid upholstered chair supported by a hard solid floor. As you read this book, you're turning pages that are also solid, yet they can be bent. Without being much aware of it, you are inhaling oxygen, a clear and formless gas, and exhaling carbon dioxide, another gas. You may be sipping coffee, mineral water, or wine.

Why does the world contain solids, liquids, and gases, and how do they differ? What causes the variety even within a given category? Such fundamental questions engaged the earliest thinkers who contemplated the physical world. Now that modern practitioners of physics, chemistry, and allied sciences can explain much about matter in its pure states of solid, liquid, and gas, they are examining it in more complicated forms, including foam.

Unlike the theory of relativity, a lightning-stroke of an idea born in Albert Einstein's unique mind, there is no single theory of foam. Studying foam requires varied scientific tools and encompasses many viewpoints. The basics of foam, however, began to be established in classical times, and were well-known by the end of the nineteenth century.

The Greek thinkers of the fifth and sixth centuries B.C.E. had an idealized view of the world. To them, everything in the universe arose from a single substance, water or perhaps air. In the fifth century B.C.E., the Greek philosopher Empedocles of Sicily proposed instead that the world is made of four elements: earth, air, fire, and water. His genius lay in choosing the four to represent varied physical properties; and in his realization that when properly combined, they could describe even subtle workings of reality. According to Empedocles, the human eye sees as water modifies fire, that is, as the fluid within the eyeball affects rays of light; in his scheme, the bones of the skeleton are made from earth, water, and fire in the ratio 1:1:2. These categories still have meaning; we now associate the elements earth, water, and air with their modern equivalents: solids, liquids, and gases.

As science continued to develop long after Empedocles' time, scientists analyzed the universe by breaking it into progressively smaller bits: first atoms and molecules; then subatomic protons, neutrons, and electrons; and then their constituents, all the way down to quarks. To grasp the world around us, we need to understand how the atomic building blocks relate to the materials around us. The connections are easiest to trace for pure crystalline solids such as salt and diamonds, whose atoms are arranged in specific patterns that repeat over and over, as a huge hotel is built up from the repeating units of its rooms. However, the real world includes clay, wood, and other noncrystalline solids. It also includes liquids, which are harder to understand than solids. And most complicated of all, there are the combinations of solids, liquids, and gases.

SOFT MATTER

Solids, liquids, and gases are combined in foams; in emulsions, where bubbles of one liquid float in another without mixing (milk, which is bubbles of fat suspended in water, is an example); and in colloids, where tiny specks of a solid substance are distributed throughout another material, typically a liquid (gelatin is an example). In each type, and unlike the orderly geometry of a crystal, the inclusions are arranged randomly in the surrounding medium. That's why these materials tend to have free-form shapes that are easily deformed, so that a bowl of gelatin sets into the shape of its container. For this reason, foam and the other combined systems are called "soft" matter—neither flowing freely like a true liquid nor taking on the hard definite shape of a rigid solid like a diamond.

Empedocles may have been the first to grasp that new properties result when substances are mingled, as his formula for bone suggests. That principle is true for any foam, such as soapsuds. It is neither fully liquid nor completely gaseous; it flows differently from the first and does not dissipate like the second. Its components are stable, yet it lives only a short while. It is made from clear air and water, yet it is opaque. And while neither air nor water sticks to the hand in any great quantity, if you scoop soapsuds onto the palm of your hand, and turn the hand over, the suds remain in place. In most ways, a foam is totally unlike the substances that make it up.

There are varied ways to make foam: by generating gas within a liquid, as when the bubbles are put into champagne; by freeing gas that was held under pressure within a liquid, as when soda spurts from an opened can; or by mixing gas or vapor into a liquid, as when air is beaten into egg whites to make a meringue, or hissing steam froths milk for a cappuccino. In the latter case, bubbles stream from the nozzle and float upward like hot-air balloons, eventually reaching the surface to form densely packed layers on top of the milk. While the foam grows, it also begins to decay, as its oldest bubbles die.

Most foams are short-lived and must be examined on the fly. Even in a long-lived foam, it is difficult to register all the bubbles as they change. And we are ignorant about how a foam flows, which depends on how its bubbles interact; do they distort each other as they move and collide, or do they tumble over each other like rocks rolling downhill? Simple direct observation can reveal several basic facts about foams, including:

- They contain bubbles of gas within a liquid or solid.
- Liquid foams tend to be white, are usually ephemeral, and move differently from a pure gas or pure liquid.
- Foams formed within solids, such as the bubbles within risen bread, generally begin as liquid foams.

After observation has yielded as much as it can, experiments and analysis are required for deeper understanding. Nowadays we think of science as requiring giant space-going telescopes and enormous particle accelerators costing billions of dollars, but it is possible to study the basics of foam at a much lower cost. In fact, I have constructed a foam laboratory that could not be cheaper—it cost nothing at all—and is remarkably informative.

To show the students in my astronomy class the foamy arrangement of galaxies in the universe, I made a simple model of the cosmos: a small transparent plastic bottle that had once held spring water, half filled with ordinary tap water and a dash of liquid soap. To demonstrate how galaxies are strewn across space, simply shake the bottle. The clear liquid turns into a mass of bubbles, each surrounded by soapy water. Stretch your mind by scaling the foam to cosmic size, so that each bubble becomes a stupendous volume hundreds of millions of light-years across. (A light-year is the distance a ray of light travels in a year, about 6 trillion miles.) Now imagine that the air within each monstrous bubble is replaced by empty space, and that the soapy water between the bubbles is replaced by billions of galaxies, each with its multitude of stars.

This plastic universe works equally well as a bubble laboratory that displays how earthly foam is born, matures, and dies. First, the laboratory demonstrates that foam requires more than pure gas and pure liquid. Omit the soap, and bubbles still form when you shake the bottle; but they die the instant you stop shaking, too unstable to give even a short-lived froth. Only by adding that bit of liquid soap, and shaking well, do you get a lingering foam. As you shake, you see a radical transformation. Transparent air and water are transmuted into an opaque white mass. The slug of liquid you feel sloshing back and forth changes as it becomes well mingled with air, replaced by a mass that hardly moves as you shake and which feels . . . well, frothier. It could not be clearer that a foam is utterly different from its constituents.

A close look through a magnifying glass shows that the foam contains bubbles of various sizes, although most are tiny, a small fraction of an inch across. They are piled like cannonballs on a courthouse lawn, but in a disorderly heap rather than a perfect stack. Each is a sphere, surrounded by liquid that isolates it from its neighbors. This is a wet foam, meaning it is more liquid than gas. If you put the bottle down, and wait, the foam changes. Slowly, the proportion of liquid decreases as the water between bubbles drains downward under the pull of gravity. It is easy to see that this is happening, because a layer of clear water appears and gradually deepens beneath the foam.

CASTLES OF AIR

As the drainage continues, the wet foam becomes something more complex: a dry foam, with more gas than liquid. The walls between adjoining bubbles become very thin; instead of each sphere floating serenely in its liquid cocoon, bubbles press against each other. In some places, the wall is breached, and two bubbles join into one. Also, smaller bubbles contain air at higher pressure, which moves through their walls into the larger bubbles. Both processes coarsen the foam, giving it more big bubbles as it ages. The shape of the bubbles also changes. Instead of spheres, adjoining bubbles distort each other into polyhedrons, three-dimensional bodies with flat or gently curved faces where they abut. Each resembles a soccer ball or a geodesic dome, except that its facets are random in size, shape, and orientation. This stage takes hours to achieve, and lasts even longer. Only after several days does the foam finally die, as the last few polyhedral cells vanish, leaving the bottle as it began, half full of soapy water.

The polyhedral stage is worth the wait, for it is an intricate and airy castle of bubbles. They come in various sizes, up to an inch across. Each nestles perfectly among its neighbors, like an irregular piece of fieldstone carefully fitted into a chimney by a master mason. With the water mostly gone, the films defining the bubbles are only micrometers thick (a micrometer is a millionth of a meter) and show rainbows of color that come from the interference of light waves within the films. Unlike the repetitive geometry of a crystalline material such as salt, the foamy structure lacks any obvious regularity. It displays irregular faces with anywhere from three to nine edges. Those that happen to have six edges are reminiscent of the hexagonal cells in a honeycomb, but the perfect symmetry of a true honeycomb is a far cry from the bewildering complexity of the foam.

The first steps toward untangling this singular structure were made by the nineteenth-century Belgian physicist Joseph Antoine Ferdinand Plateau, the most influential early foam scientist. He derived laws for the geometry of a foam that still

A foam containing polyhedral cells.

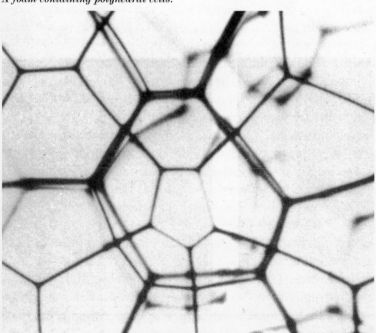

stand today. Ironically, this student of foam geometry was blind for much of his life, a result of his early research in optics, during which he gazed directly at the Sun. While Plateau could still see, he began to study soap films stretched over wire frames; with the aid of relatives and colleagues, he continued to do so even after losing his vision.

Plateau developed a mixture of soap, water, and glycerin that gave films and bubbles that survived for up to eighteen hours, allowing him to make careful observations. Eventually he evolved a set of universal laws, Plateau's rules, that impose some order on the mass of bubbles in a foam, and that can be confirmed in the plastic bubble laboratory. If you carefully examine the films forming the walls of the bubbles, you can conclude, as Plateau did, that: (1) only three films—no more, no less—ever meet to form the edge of a bubble; (2) any two adjacent films of these three always meet at an angle of 120 degrees; (3) exactly four edges of bubbles—again, no more, no less—ever come together to meet at a point.

Plateau's century-old rules are a triumph of observation and experiment. They are never violated, no matter how intricate the foam. But why are these fundamental rules true, and can we go beyond them to more fully describe a foam and its oddly shaped bubbles?

SURFACE TENSION

The remarkable geometry of a foam is sculpted by natural forces. To understand these forces, we must go back to early studies of gases and liquids, air and water; and of their meeting place, the bubble. Water is a good starting point. Compared to other liquids that support foam, such as milk or egg whites, water has a simple microscopic structure. The familiar symbol for its molecule, H_2O, represents two atoms of hydrogen and one of oxygen, bound by electrical forces into a boomeranglike shape, with the oxygen at its center and a hydrogen defining each arm.

Water is a collection of such minute boomerangs. Even when water sits motionless in a beaker, bowl, or pond, all of its molecules—an astronomically huge number—are in constant motion. These molecules attract each other (if they did not, water would fly from an open container like a gas), and that fact goes far in explaining how water supports bubbles and foam. They form because of the force called surface tension, which is due to molecular attraction. This molecular understanding is a relatively recent development, but surface tension has been known for centuries. Leonardo da Vinci entered his careful observation of surface tension's effects into his notebook, around 1508. Describing water as it drips slowly from a source, he wrote: "That water may have tenacity and cohesion together is quite clearly shown . . . where the drop [of water] before it falls becomes elongated as possible, until the weight of the drop renders the tenacity by which it is suspended so thin that this tenacity, overcome by the excessive weight, suddenly yields and breaks."

The modern picture of molecules reveals why water acts in this way. Think of the many H_2O molecules making up the water in a beaker. Forget for the moment that each molecule has a specific boomeranglike shape, and think of it only as a bit of matter attracted by identical bits of matter all around it. Now imagine trying to pluck a single molecule from the beaker with extremely fine tweezers. If your tweezers close around a target deep within the water, that molecule can be easily pulled loose. The reason is that although all the neighboring molecules attract the target, since they surround it in three dimensions— above and below, left and right, front and back—all their pulls average out to zero.

Now clamp your tweezers around a molecule at the top surface of the water. As you try to pluck it loose, the molecules be-

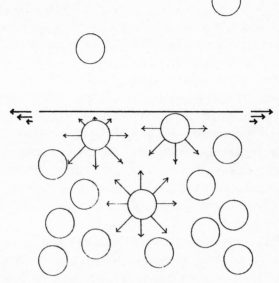

Molecular forces illustrating surface tension in water.

low the target draw it down toward the bulk of the water. With no water above, no molecular force opposes the downward pull, which you feel as resistance to your efforts. This is surface tension—the force that pulls molecules on the open surface of a liquid toward the interior of the liquid, making a drop of water behave as if it were coated with a taut elastic skin. Surface tension appears throughout nature. It plays a role in how water rises from the roots to the top of a plant; it enables the insect called the water strider to literally walk on water, supporting the insect's feet as if they treaded a miniature trampoline. And surface tension is responsible for the formation of drops of water and bubbles of gas within water.

After da Vinci's time, other researchers examined surface tension and bubbles. British physicist Charles Vernon Boys, a particularly deft experimenter, became intrigued by bubbles in the late 1880s. His genius lay in demonstrating their properties so lucidly that schoolchildren could follow his lectures, which were published in 1890 as the classic work *Soap Bubbles, Their Colours and the Forces Which Mould Them*. Boys employed the most minimal of equipment, such as wire frames, gas flames, and thin fibers made from quartz, a pure form of glass. (Boys first used such fibers to measure minute gravitational forces when he determined the Earth's density. He made them in a typically clever manner: He would dip a crossbow bolt into molten quartz and shoot the bolt into the air, drawing along a liquid tail that solidified into an exquisitely thin thread. Although the thread might break when the bolt landed, this method provided the short bits of fiber that were all he required.)

Boys had a marvelous way of using these simple implements to make abstract properties of bubbles immediately apparent. For instance, he showed that water held within a thin rubber bag behaves like a drop of water under the influence of surface tension, giving convincing concreteness to the effect. And drawing on the late-nineteenth-century knowledge of light and optics, which was well advanced, Boys and his fellow scientists knew that a wet foam looks white because rays of light are deflected or scattered by its bubbles, and that a dry foam shows colors because light waves interfere within its thin films.

THE MINIMIZING PRINCIPLE

Boys and his colleagues also understood the role of energy in bubbles and foam, a powerful insight that explains why isolated bubbles are perfect globes. It had been established much earlier that any physical system is most stable at its lowest energy and acts to reach that equilibrium if at all possible. This behavior is so widespread and well-known in physics that it does not go by a special name, but I'll call it the minimizing principle for easy reference. The principle explains why a ball rolls downhill; why a stretched spring returns to its relaxed state when it is released; and why a laser emits light. In each case, the action happens as the physical system returns to a lower energy after being raised to a higher one. And the minimizing principle is the reason that drops of water and bubbles take on their particular shapes.

It takes energy to maintain a drop of water, and the larger its surface area, the more energy. So the minimizing principle requires that surface tension always acts to minimize area. For a given amount of liquid, the shape with the smallest surface is a sphere, and surface tension tries to pull a drop of water into a perfect globe, although that cannot always be completed. A raindrop forms a bead on a newly waxed automobile, but not a full sphere, since part of the drop clings to the vehicle so strongly that surface tension cannot pull it away. Water dripping slowly from a tap forms a more complete sphere, although elongated by gravity, before it breaks free and falls. Under zero gravity, however, as encountered by astronauts, water forms perfectly round droplets. (Joseph Plateau did not need NASA to demonstrate those ideal spheres. He made bubbles of oil within a mixture of water and alcohol which he had carefully measured out to obtain exactly the same density as the oil; that effectively canceled the effects of gravity, giving beautiful round globes.)

The ideas of surface tension and minimizing principle apply equally well to a gas bubble inside a liquid, which consists of an elastic skin enclosing a volume of gas, like a balloon. Surface tension provides the skin, but something else is needed to make the bubble truly robust. Such a material is called a surfactant (an acronym for *surface-active agent*), and soap is a prime example. Its molecules, which are released when soap dissolves in water, act to modify surface tension so it can vary across the curvature of a bubble. This allows the bubble to adjust to gravity or other forces that would otherwise destroy it. In addition, the soap molecules interlock with water molecules, thereby strengthening the skin around the bubble; to put it another way, they create a skin from which water drains only slowly, extending the lifetime of the skin and the bubble. Make a large number of such bubbles, with enough surfactant to make them hardy, and you have created a foam.

Soap is not the only surfactant. Seawater contains a rich stew of molecules that play the same role. That is why the sea sustains foam, whereas freshwater does not, unless it contains additional molecules. If these come from polluting compounds, the result may be a foamy scum, the mark of contaminated water. Many edible foams, such as soufflés and whipped cream, are based on egg whites or milk products, whose complex molecules are surfactants. Some surfactants make foams that are resilient, meaning if a rift develops in the film around a bubble, the film flows so as to heal the opening. Others give plastic foams, meaning that liquid and surfactant form especially strong films that allow the bubbles to survive for hours.

No matter what the surfactant, the minimizing principle always applies. This makes it possible to use soap films to answer difficult mathematical questions called minimal area problems. These arise in mechanics, optics, and the theory of relativity, but the simplest of the type has practical meaning. If you want to build a highway linking a certain number of towns, what configuration requires the least length of road (and therefore the lowest cost) while ensuring that one can drive between any pair of towns? For two towns, the answer is obvious: a straight road. But for an arbitrary number in an arbitrary pattern, it is surprisingly elusive. In *The Science of Soap Films and Soap Bubbles*, British physicist Cyril Isenburg presents a set of striking photographs showing how a soap film automatically indicates the correct route when it clings to a small scale model giving the location of the towns. Other photographs in the series show how a film naturally takes on minimal area under intricate three-dimensional constraints, such as being required to touch every edge of a cube.

The powerful minimizing principle is more than a guide to the shape of an individual soap film; it also requires that the bubbles in an entire foam adopt the most stable shapes. The connection between Joseph Plateau's rules and the minimizing principle was not proved until nearly a century after Plateau's death. That tour de force was carried out at Princeton University in 1973 by American mathematician Jean Taylor, who showed that Plateau's rules could be directly derived from the minimizing principle.

LORD KELVIN'S BEDSPRING

The application of the principle to a foam leads to a classic problem in mathematics. Each bubble within a foam must take a shape that gives minimal surface area and must also be consistent with the constraining presence of its neighbors. If the bubbles are isolated, as in a wet foam, that shape is a sphere. But in a dry foam, where only thin films separate the bubbles, they must fit together to completely fill the volume of the foam, which cannot be done with spheres. A more complex shape is needed, which turns out to be hard to define mathematically even for a foam simplified and idealized, with all bubbles the same size. This particular pattern of nature is a long-standing enigma whose solution is still elusive. Its answer is important in foam science, because without knowing the shape of the bubbles, it is difficult to predict the properties of a foam.

One of the first to consider the problem was the distinguished British physicist William Thomson, Lord Kelvin. He invented the idea of absolute zero, helped develop the second law of thermodynamics, and contributed to the laying of a transatlantic telegraph cable. In 1877, drawing on shapes known for iron crystals, Kelvin theorized that the bubble with minimum area was a strange-looking figure with six square and eight hexagonal faces. He modeled this remarkable shape, called a *tetrakaidecahedron* ("fourteen faces" in Greek), out of wire, producing what came to be known as *Kelvin's bedspring*. This was long taken as the definitive minimum-area shape, but had never been tested in an actual foam. Finally, over fifty years after Kelvin's conjecture, the American botanist Edwin Matzke constructed, by hand, a foam of thousands of identical bubbles to discover how cells pack together inside living organisms. However, he failed to find even one of these Kelvin shapes within the mass of bubbles. Although Matzke's experiment was inconclusive, it did cast doubt on whether Kelvin's shape is truly optimum. Even now, the optimum shape of bubbles in a foam is not definitively known.

Early foam scientists used experiments coupled with ideas such as surface tension and the minimizing principle to establish basic rules about bubbles and foam. These rules do not explain everything, such as the shape of the bubbles in foam, or the extraordinary behavior of a foam under stress, that is, when it is pushed or pulled. Under gentle force, a foam holds its shape like a solid, whereas under greater stress, it flows like a liquid. That is why whipped cream stands firm under its own slight weight but spreads easily when urged by a spatula. These and other characteristics of foam, such as its lifetime, raise scientific questions and are important for human needs. Longevity, for instance, may be desirable in whipped cream but not in soapsuds polluting a natural waterway. And so after the understanding that had developed through the nineteenth century, twentieth-century scientists set about finding new techniques to further enlighten us about foam.

EXAMINING FOAM

IMAGERS, LASERS, AND COMPUTERS

Scientists now go far beyond the classical simplicity of wire frames and quartz fibers, using modern tools to examine foam. These studies expand our fundamental knowledge of matter, bearing on such questions as how it is distributed in the universe. Closer to home, deeper knowledge will make it possible to design useful new foams, such as insulating materials for homes and vehicles, and fire-fighting foams to snuff out flames. The modern study of foam and its bubbles employs tools as varied as lasers, space science, and sophisticated imaging methods.

IMAGING FOAM

The first step in the study of foam is to examine its bubbles, which is not easy because light does not readily penetrate a foam. One ingenious approach modernizes a simple technique used by Charles Vernon Boys. He made a cell with transparent glass sides, like an aquarium tank except that it measured a mere half inch from front to back. When he poured in a solution of soapsuds, only a layer or two of bubbles could fit into that narrow space. As these were not obscured by other layers, Boys could easily examine their size and shape. Today's researchers study foam by combining a glass cell with modern office technology. The cell is filled with foam and placed horizontally on an ordinary copying machine. A rapid-fire sequence of images is made, each a pic-

Charles Vernon Boys

ture of the bubbles at a different time, which shows the evolution of the foam.

Another innovation takes the glass cell a step farther. Scientists at the University of Paris have come up with a kind of magnetic soapsuds, which allows them to explore the dynamic
behavior of a foam. Instead of studying bubbles of air formed in
water containing molecules of soap, they study bubbles of oil
(which, like air, is less dense than water) formed in water laced
with tiny magnetic particles. The particles link together
through their magnetic forces, as soap molecules do through
atomic forces, to act as surfactants. When this mixture is placed
between glass plates, it forms the familiar polyhedral cells seen
in soapsuds. These cells change as the researchers bring a mag-
net near, providing a simple way to study the dynamics of a
foam under varying conditions.

Boys's original technique and its updated versions, however,
suffer from the same lack: They examine only a thin slice
through a foam. This two-dimensional look at a complex three-
dimensional structure is like trying to picture Michelangelo's
David from a single cross section through the statue. Another
problem is that the walls confining the bubbles create forces
that distort them, so a layer held between sheets of glass is not
even a true slice through the heart of a foam. It would be far
better to analyze the foam in its full three-dimensionality, but
that is difficult. One reason is the sheer volume of information;
the other is the difficulty of seeing through a foam.

A wet foam appears opaque not because it absorbs light, like
a black object, but because it scatters light, which is also the
reason a foam looks white. After a ray of light enters plain water, it continues in an undisturbed straight line, but in a foam,
the ray soon encounters a bubble, where part of the light is reflected and part continues through the bubble. This produces
rays traveling in directions drastically different from the original line of motion. The new rays encounter other bubbles,

which again alters their directions; and so light rays wend through a foam like balls in a three-dimensional pinball machine, bouncing off bumper after bumper to trace complex, nearly unpredictable trajectories. Although some light may survive to emerge from the far side of the foam, the original arrangement of rays that defines an image is lost through the multiple scatterings. It is impossible to see a meaningful scene through foam; under ordinary white light, all that comes through is an undifferentiated white glow.

Even though all semblance of an image is lost through scattering in foam, scientists exploit the process by shining a laser beam through the foam and examining what emerges. The exiting light has encountered many bubbles in its travels and carries information about the deep interior that can be recovered. This optical probe, called diffusing-wave spectroscopy (abbreviated DWS) because it measures scattered or diffused light, was first applied to a foam in 1991 by physicists David Weitz, Douglas Durian, and David Pine, then working at the Exxon Research and Engineering Company. To establish the method, the researchers examined a commercial foam, Gillette shaving cream, fresh from the can. They put some into a glass cell, shone a laser beam (which happened to be blue) through one face of the cell, and measured the fraction of the laser light that emerged through the opposite face after traversing the foam.

The researchers then derived a simple and powerful relation between the emerging light and the properties of the foam: The amount of light that penetrates a foam changes along with the average size of the bubbles. Knowing this, they could actually watch the shaving cream coarsen; as time went on, the foam transmitted more light, showing that the bubbles became bigger. When the foam was 200 minutes old, for instance, the average bubble was twice as large as when it was 40 minutes old.

Unlike Boys's fish tank, this method examined coarsening throughout the foam, including its deep interior, not just in a two-dimensional slice. It also revealed something new. The experimenters noticed that the intensity of the transmitted light fluctuated, indicating that something was rapidly changing in the foam. They carefully examined the bubbles through a microscope and eventually observed *rearrangement events,* where several neighboring bubbles would suddenly and simultaneously shift their positions. The researchers surmised that as coarsening continued and the bubbles changed in size, stress developed, eventually reaching a value such that it was easier for the bubbles to move than to take on new shapes.

Although the DWS method is valuable, what is really needed is a way to measure three-dimensional foam data. Magnetic resonance imaging (MRI), the diagnostic method that allows physicians to look inside a living human body, makes this possible by completely pervading a foam. MRI grew out of another technique, nuclear magnetic resonance (NMR), developed in the 1930s and 1940s to explore materials through the magnetic properties of atomic nuclei. The material is placed in a strong magnetic field and exposed to high-frequency radio waves. The magnetism makes the atomic nuclei within the material rotate at a certain rate, according to the same laws that make a compass needle deflect in the Earth's magnetic field. If the rate of rotation matches the frequency of the radio waves, the nuclei selectively absorb energy from the waves (the "resonance" part of "nuclear magnetic resonance"), which allows their properties to be determined.

In the early 1970s, researchers began turning NMR into MRI, which has since evolved into a routine medical procedure. As in NMR, the human subject is immersed in a strong magnetic field and exposed to weak radio waves, both biologically harmless. This makes it possible to measure the locations and concentrations of hydrogen atoms, as part of the water prevalent in the body. A computer turns that information into images of structures inside the body.

MRI looks equally deeply into an opaque foam if the foam contains magnetically responsive nuclei like the hydrogen in water. The first three-dimensional portrait of a foam was made in 1995 at the University of Notre Dame. Researchers, using a commercial MRI machine to examine a foam made of gelatin mixed with water, produced a striking three-dimensional image of its multitude of bubbles. Since then, MRI has been used to determine the sizes of the bubbles as a foam ages, providing a tool to examine changing foam. But because the imaging takes so much time, MRI cannot follow rapid change such as that seen in a flowing foam. Only optical methods like DWS respond quickly enough.

THE DYNAMICS OF FOAM

Douglas Durian, one of the team who established the DWS method, extended the technique to study foam in motion. After working at Exxon, Durian took a faculty position at UCLA; there, in 1995, he and his student Anthony Gopal used a variant of Boys's fish tank to examine a moving foam. Placing shaving cream between two glass plates, they slid one plate over the other. That smeared the cream along the direction of motion, just as a spatula spreads whipped cream, a process called shearing. They monitored the moving foam by DWS and found they could relate the degree of shear to what Durian calls the "complicated collective dance" of rearrangement events. But DWS and other optical methods cannot reveal everything about how foam moves. There is a basic constraint on the examination of foam in motion: The force of gravity pulls its liquid component downward, shortening the lifetime of the foam. Gravity also makes it necessary to confine foam in a container for study, so it does not spread out. Yet the container itself alters the behavior of the foam.

Douglas Durian

One solution, however, defies gravity. Glynn Holt of Boston University uses acoustic levitation to keep a drop of foam floating in air, thus avoiding the distortions that arise from contact with a container. A transducer—a small audio speaker-like device that vibrates when voltage is applied—is pointed straight up at an equally small metal reflector mounted an inch or so above and facing straight down. There is nothing to hear when the transducer is switched on because it works at frequencies above human range; but the transducer generates sound waves that bounce back and forth between it and the reflector, until the space between contains enough acoustic power to move a drop of foam.

With everything properly adjusted, a bit of foam is carefully squeezed out of a hypodermic needle into the space between transducer and reflector. The sound waves push and pull it just right to magically hover in midair, while a video camera puts an image of the drop on a monitor. By displaying the complex vibrational behavior of the drop, the monitor gives information about the dynamic mechanical properties of foam. The acoustic technique, however, does not do away with the drainage of the liquid under gravity. So Holt, who was trained by NASA as a payload specialist for flights into space, wants to examine foam under true low-gravity conditions, aboard the international space station now being built in Earth orbit.

VIRTUAL FOAM

Computers allow us to sidestep some laboratory limitations and calculate the properties of a foam, or create a virtual foam that simulates a real one.

Computational foam science is helping enormously in finding the shape of the bubbles in a foam. Researchers continue to seek forms with areas less than the area of the tetrakaidecahedron (which Lord Kelvin proposed in the nineteenth century) in order to discover the shape with minimal area. The hunt continues partly to find the answer to an intriguing mathematical puzzle, but there is also a reason of bedrock importance for foam science and technology. Without knowing the shapes of the close-packed bubbles in a foam, it is impossible to accurately predict its mechanical and dynamic properties such as how it flows or yields to stress.

Mathematical physicist Denis Weaire, of Trinity College, Dublin, with his colleague Robert Phelan, recently found a likely candidate by using a computer to find and test new shapes. The process relied on a well-known computer program called the Surface Evolver. It uses triangular tiles to create a three-dimensional surface of any desired shape and complexity, and then reshapes the surface so as to minimize its area. This procedure would be unbearably slow and difficult by any other method.

Weaire had once sent a student "on safari" through a foam, photographing its bubbles as part of an undergraduate research project. The photographs had shown traces of a structure similar to the shape produced by the computer program. This Weaire-Phelan structure is more complex than Kelvin's bedspring. It combines two basic shapes: One has twelve curved pentagonal facets, and the other adds two hexagons to the twelve pentagons for a total of fourteen faces. The power of Lord Kelvin's insight is illustrated by the fact that the area of the new structure, derived a century later with the aid of a computer, is smaller by less than 1 percent—but smaller it is. There is as yet no assurance, however, that this is the structure with the very smallest area, or that it truly exists as calculated. Its discoverers are continuing their hunt through real foams to confirm their computer result.

Computers can also simulate the dynamics of a foam. The first step in simulating a foam—or any physical system, from a sand dune to a star—is to identify its key elements, typically its smallest parts and the forces acting on them. The computer calculates how the forces affect each basic element and combines their separate motions into a description of the entire system. The calculated result of a computer simulation is usually put into pictorial form for easy comprehension. Such pictures can be seductive, and researchers must interpret them with care. A simulation can be worse than useless if it is based on poor assumptions, and there is always the danger of omitting some important physical effect. Applied with appropriate wariness, however, the method can be highly informative.

Douglas Durian and his wife, chemist Andrea Liu, recently ran a computer simulation to explore moving foam. Its elements are the bubbles themselves, represented by the mathematical equations for spheres; the liquid, which makes itself felt by the friction it exerts on the bubbles; and the forces between the bubbles, which model how they press on each other. Even with a big computer, the calculations are demanding, and this virtual foam was limited to several hundred bubbles moving in two dimensions. The computer calculated the size and position of the bubbles as they responded to different levels of stress, and drew a picture of the foam in each case. According to Durian, these snapshots show that avalanchelike events occur when a foam flows. As the foam shifts, large numbers of tightly packed bubbles suddenly snap from one configuration into another, like rocks on a hillside frantically tumbling over each other as an avalanche roars down the slope.

THE BUBBLE CHAMBER

The next step is to correlate the results of computer simulations and optical experiments and MRI measurements and a dozen other approaches until a consensus emerges among those who study foam; only then can we truly feel we are beginning to understand its dynamic and static properties. Although there is still a lot to be learned about bubbles and foams, we know enough to use them as scientific tools and sources of new science. Bubbles, for instance, have made it possible to observe the behavior of tiny elementary particles.

No microscope can magnify a proton or a quark to visible size; they can be seen only indirectly, through the tracks they leave. These tracks can be observed in the device called the bubble chamber, invented in 1952 by physicist Donald Glaser; he received a Nobel prize for the idea, which supposedly dawned as he watched bubbles in a mug of beer. A transparent container is filled with extremely cold liquid hydrogen (other liquids also work), held under particular conditions of temperature and pressure. An elementary particle that enters the chamber causes the hydrogen to boil along its track, leaving a wake of bubbles that can be photographed and measured. The device has been essential for fundamental physics. In 1973, a team of researchers examined over a million bubble-chamber photographs and found evidence for the breakthrough "electroweak" theory, a major step in understanding how nuclear and electromagnetic forces are related.

BUBBLES OF ENERGY

One other area of research may lead to a dynamic foam that generates clean and limitless energy. That is the phenomenon of sonoluminescence, in which a very small gas bubble, floating in water, converts sound waves into a burst of light. The sound is generated at a frequency inaudible to humans, but it is as intense as that from a shrieking smoke alarm. In the water, the sound waves create alternating pulses of high and low pressure that make the bubble contract and expand in the same rhythm. At the bubble's maximum size of about fifty micrometers, the gas it contains exerts hardly any outward pressure. The surrounding water pressure then drives the bubble violently inward, shrinking it fifty times or more. At that point, the bubble emits a flash of light that comes and goes in mere trillionths of a second, as determined by physicist Seth Putterman at the University of California, Los Angeles. The flash repeats once every cycle of the sound waves, with such faultless regularity that sonoluminescing bubbles are used to time the motion of elementary particles moving near the speed of light.

The startling feature is that the light lies mostly in the invisible ultraviolet part of the spectrum. According to the laws of radiation, that means this tiny collapsing bubble, buried deep within a liquid, somehow reaches temperatures up to 100,000 degrees Celsius (180,000 degrees Fahrenheit), far hotter than the surface of the Sun.

The leading theory to explain sonoluminescence is that the rapid implosion creates a shock wave, a moving zone of high pressure where the gas in the bubble is greatly compressed, enormously raising its temperature. In support of this view, measurements in Putterman's laboratory show that the bubble contracts at greater than Mach 4, that is, four times the speed of sound or thousands of miles per hour. Even more impressive is the acceleration of the bubble as it expands after compression, which is several billion times that of gravity. That would bring the bubble wall to the speed of light in a fraction of a millisecond, except it is not sustained for nearly that long a time.

What gives sonoluminescence special weight is the speculation that it might be used to induce thermonuclear fusion, the process that brings together atoms of hydrogen to form helium in the heart of our Sun or any star, releasing enormous energy that makes the star glow for millions or billions of years. Fusion is a clean and efficient form of energy production but requires temperatures of millions of degrees to force the hydrogen nuclei together. This and other difficulties have kept fusion from reaching useful status on Earth, despite large sums of money spent in the last decades to build experimental fusion machines.

The high temperature in a sonoluminescing bubble suggests that one of these violently imploding spheres, filled with the proper form of hydrogen, might be a minute furnace for thermonuclear fusion. No one has yet measured temperatures remotely high enough, but some researchers believe that millions of degrees could be reached by changing how the sound waves are applied. This suggests the stimulating image of a host of madly oscillating sonoluminescent bubbles, each providing its mite of thermonuclear energy, the most dynamic foam one could possibly imagine.

Research is under way to determine whether this energetic foam will ever be realized. Putterman is planning experiments in which the sonoluminescing bubble is surrounded by several detectors of light rather than the single one now used. That will give refined information about the shape of the spot of light, an important clue to its origin. Eventually we may be able to predict whether a foam of sonoluminescing bubbles can produce temperatures high enough for fusion, and then begin to build such a generator.

SURPRISES FROM FOAM RESEARCH

Learning how foam moves and coarsens will make it possible to design foams with specific characteristics for applications like fire fighting, or to minimize the environmental impact of foamy industrial by-products. But some benefits are unexpected. Douglas Durian uses DWS and computer simulation to examine how sand flows, a surprisingly ill-understood phenomenon. The same computer program that simulates foam simulates sand by making the individual spheres rigid like sand grains rather than soft like bubbles. This may illuminate problems of ecological and geological scale: how sand dunes form and move, how earthquake faults propagate through the Earth's crust.

The abstract mathematics of foam carried out by Jean Taylor, who solved the problem of connecting Plateau's laws to the minimizing principle, also displays unexpected benefits. To attack the problem, it became necessary to develop a new mathematics of shape called geometric measure theory. This is a potent tool with wide application in the real world: designing intricate electronic chips so as to avoid the possibility of short circuits, ensuring that liquid fuel in a NASA space shuttle reaches the outlets in the fuel tanks no matter how the shuttle maneuvers, and making stronger turbine blades for jet engines.

We now probe foam with lasers, suspend it in space, and simulate it with computers. At one level, this is fundamental research that simply aims to better understand the world around us. At another, it underpins the development of real foams with a tremendous range of human benefits—some mundane, such as improved shaving cream; some enormously important, such as a power-producing sonoluminescent foam.

While scientists work to deepen their understanding of foam, others make and use foam by drawing on lore that comes from long experience. The next chapter relates how people used the attributes of foam to bake bread and brew beer beginning long before there was a science of foam. We still enjoy these classic edible foams, as well as newer types such as champagne, and whipped cream that billows out of a can.

EDIBLE FOAM

BREAD, BEER, AND CAPPUCCINO

For all the progress we have made in eating for health while eating well, from counting calories to watching cholesterol, we take in surprising quantities of nonnutritious air and carbon dioxide among all those substantial proteins and good vitamins. Much of what we ingest is in foamy form, where the nutritious part is combined with gas, which lends no sustenance but enhances the texture, appearance, and pure enjoyment of what we eat and drink.

There are so many edible foams that they could make a tasty and satisfying meal, starting with a frothy soup, progressing to a cheese soufflé and chocolate mousse and including good things to drink. There are also the foams we eat or drink between meals or to satisfy a sweet tooth: ice cream, marshmallows, milk shakes, and that New York City delicacy, the egg cream.

The cuisines of most cultures include foamy food and drink. Yorkshire pudding, that staple of traditional British cooking, has a foamy nature; the Emmentaler and Gruyère cheeses of Switzerland are foamlike in containing bubbles of gas within a solid matrix; and the sparkling Italian wine Asti Spumante is named after *spuma,* the Italian word for foam. Similarly, many cuisines include a form of leavened, or raised, bread, which is a solidified foam filled with gas bubbles. Then there is whipped cream, the lavish grace note of French and Viennese pastry.

French cuisine owes a great deal to foam. In 1720, when a Swiss chef named Gasparini baked egg whites whipped with air into the first meringue (named after his hometown of Meiringen), he was contributing to a long line of edible foams. A hundred years later, the mousse (which means "foam" in French) and the soufflé, as well as the meringue, were being counted among the glories of French cooking, as recognized in the gastronomical writings of Jean-Anthelme Brillat-Savarin.

Each of these edible foams begins as gas mixed into liquid. For some, the gas is a by-product; for others, it is deliberately added. In bread and beer, the liquid is water, laden with complex molecules that behave as surfactants, and the gas is carbon dioxide, or CO_2, a tasteless and colorless compound that is harmless when ingested or inhaled. This gas is a bonus from the process of fermentation, in which yeast turns sugar into alcohol in both bread and beer. A meringue is built from a different recipe: The liquid is egg white, which basically is water filled with protein molecules that hold the foam together, and the gas is air that has been mechanically beaten into the liquid. In most foamy foods the liquid foam is cooked or baked (whipped cream is an exception), which alters its flavor and makes it firmer. Depending on the constituents and the degree of heating, baking can give a self-supporting but airy solid foam like a meringue or a heavy, substantial one like bread.

Not only are bread and beer closely associated with foam, but they share other similarities: Both are millennia old, and both depend on the cultivation and processing of grains, which marked human civilization's turning from hunting and gathering to an agricultural way of life. Wine is another ancient fermented foodstuff that has been with us for millennia, and although in olden times it was not necessarily made to be bubbly or foamy, it has become so in champagne and sparkling wine.

Our long experience with edible foams has taught us how to make them more foamy or less, softer or firmer, shorter- or longer-lived—all factors in their enjoyment. There are more subtle issues as well; for instance, although drinking beer would not be the same without the sensual experience of the head (one early brew was ingested through a straw, which seems a shame), too much foam can reduce the tastiness of beer. This only confirms what every beer drinker knows: There is art beyond science in brewing, including the amount of gas in the final product; and further art in how the beer is poured.

Although eating is a necessity, enjoyment is an important factor in our food choices. Another is convenience. In the modern world, where technology influences how and what we eat, increasingly our food is canned and boxed, sliced and diced; preprepared and precooked, frozen and recooked. Many foamy foods require skillful cooking that does not lend itself to mass production, and their delicate flavor and appearance vanish unless these foods are made and consumed on the spot. It is hard to imagine freezing a fluffy soufflé for later thawing, or canning a cappuccino in the hope of re-creating its foamy head and particular flavor when it is reheated.

But although some edible foams will never fit into a mass-produced, fast-food sensibility, technology has also enhanced foamy food and drink. The story of how espresso came to be available on every street corner is a saga of steam power and water pressure applied to make a perfect coffee foam. And when humanity first learned to make bread and beer and wine, it was beginning to evolve the complex technology of fermentation common to all three. Applied to different products of the Earth—wheat for bread, barley for beer, grapes for wine—it is what makes each food a foamy one.

BAKED AND BREWED

Bread is thousands of years old; we know that the Egyptians were baking leavened bread from wheat flour as long ago as 2,000 to 3,000 B.C.E. Surprisingly, beer may have been with us even longer. Tablets preserved since the days of Mesopotamia, whose cities of Babylon and Nineveh lay between the Euphrates and Tigris Rivers, show that beer was being brewed in 6,000 B.C.E., and it may be older than that. Barley may already have been a domesticated grain 10,000 years ago in the area of the Middle East called the Fertile Crescent, which extended from present-day Turkey into present-day Iran. For beer and bread, it was the connection to grains that influenced how civilization grew. Although grain was probably first gathered in the wild, it takes a stable and organized settlement to cultivate and process barley or wheat. Anthropologists and archaeologists are still debating whether bread or beer was more important in influencing human society toward agricultural forms.

The discovery of fermentation preceded any scientific understanding of chemical reactions, gases and liquids, and foam—in fact, it preceded science itself. We now understand fermentation, but do not know exactly how it first occurred to raise bread and make beer. For bread, once grain had been pounded into flour and mixed with water, airborne natural yeast might

Bread slice (left). Close-up view of bread (right).

have settled on the result and initiated fermentation. For beer, one theory is that the ancients crumbled bread into water, producing a liquid that fermented. Whatever the origins, the invisible progress of fermentation must have carried an aura of mystery in those days. There are hints of how the ancients struggled to grasp the process: the Greek root for *yeast* and the Latin root for *fermentation* both mean "boiling," suggesting that the bubbling accompanying fermentation was interpreted as a kind of cold boiling. It is not surprising that bread and beer both took on magical and religious meaning those millennia ago.

The invisible agent that made bread rise and beer froth remained unknown until it was isolated and identified as carbon dioxide in the seventeenth and eighteenth centuries. By the late nineteenth century, Louis Pasteur and others had investigated fermentation and found that it was caused by yeast, a microscopic single-cell living fungus which combines with sugar to make alcohol and carbon dioxide. In both baking and brewing, the first step is to change starch, a main component of wheat and barley, into sugar, which is then fermented to make alcohol and carbon dioxide. The fermentation that produces wine is simpler by one step, since grapes already contain a high proportion of sugar.

In bread, the alcohol produced by fermentation evaporates during baking. The carbon dioxide would also escape except for gluten, the component of wheat and rye flour that enables bread to rise. Gluten is primarily made of proteins, which are large, complex molecules composed of thousands of atoms of nitrogen, hydrogen, carbon, and oxygen formed into specific combinations called amino acids. Proteins are essential to living things, forming a large part of muscle and other types of tissue, and enhancing metabolic processes.

These molecules are also important in edible foams such as risen bread because of their shapes and how they combine with each other. The proteins in gluten link together in long chains, forming a network that holds the carbon dioxide within individual cells, which expand along with the gas during baking. (This linking tendency also makes gluten useful as an adhesive.) The heat of baking induces other chemical processes that join molecules of starch, and binds the gluten to the starch, to firm up the malleable dough. The final result, as you can see in any slice of risen bread, is a solid material punctuated with numerous cells arranged in a random pattern.

The size and distribution of the cells indicate how completely the dough has been kneaded—that is, pushed and pulled to thoroughly combine with the yeast for complete fermentation. Well-kneaded bread contains many small cells; insufficient kneading produces a coarse texture with large pockets. The amount of kneading required depends on the leavening agent. Yeast is not the only way to make bread rise, although it is the most effective. Baking powder and baking soda produce carbon dioxide by chemical reaction, and in the sourdough process, leaven is produced in the dough itself. (Some baking techniques use no leaven at all: Batter mixed vigorously with air and egg whites, or even with steam, produces fluffy baked goods like angel food cake.) The ancient Romans recognized the importance of kneading and worked dough with paddles turned by a donkey walking in a circular path. Today, bakers use power machinery to perform that essential task, and homemakers use countertop bread machines.

In prerevolutionary France, where the best bread in Europe was made, kneading was done by hand. Much of that good bread was made without adding yeast, using a method akin to the sourdough process. To make the bread rise sufficiently, French bakers spent hours of hard physical labor manipulating piles of dough. (Historian Steven Laurence Kaplan writes that the baker was said to "mount" the dough as he massively kneaded it, as if intimate sexual contact between baker and dough was the prelude to the birth of a loaf, and a symbol of bread's role in maintaining life.) The savvy eighteenth-century French consumer knew that an even dispersion of small pockets was convincing evidence that a loaf of bread had been well and truly kneaded; in modern France, this is still taken as a sign of quality in a baguette.

Each method of leavening produces its own particular taste, texture, and degree of foaminess—differences that have sometimes turned bread into a symbol for class differences. Over the centuries the most obvious distinction has expressed itself as white bread for the gentry, dark bread for peasants and workers, but softness has also carried symbolic value. By 1,000 B.C.E., for instance, the Egyptians had been baking bread for a long time and had many different types; among them, leavened bread was reserved for aristocrats, and flat unleavened bread for common folk.

A similar class distinction based on the lightness or foaminess of bread existed in seventeenth-century Paris and went on to become a matter of politics and public health. The favored bread of the rich at the time was *pain mollet,* known for its lightness and fluffiness. These qualities were achieved because the bread was made with barm, or brewer's yeast, a fermenting agent that illustrates the close ties between bread and beer. Barm is a yeasty froth that arises as beer is brewed, and can be skimmed off for use as a leaven. (The term appears in British slang as *barmy,* meaning "weak-minded, mentally lightweight, or foolish," presumably through the association with frothiness.) Ordinary bread was heavier, because it was made by the sourdough method.

In a scenario reminiscent of Jonathan Swift's *Gulliver's Travels,* where war breaks out between the kingdoms of Lilliput and Blefuscu over whether a soft-boiled egg should be broken at its big end or its small one, the two styles of bread engendered passionate differences of opinion. As contemporary comment had it, according to Kaplan, heavy ordinary bread was proper for working folk, whereas soft *pain mollet* was immoral for them because it emphasized "sensual pleasure," which could "draw the common people into error" and waste resources. There were medical objections as well. Brewer's yeast was called a "scum" that came from rotting barley and adversely affected whatever it touched. The proponents of barm were equally pungent; dismissing ordinary leavening as "coarse," they described barm as "more subtle and more penetrating," leading to a bread that was tasty and healthy. The differences came to a head in Paris in the 1660s, when physicians, statesmen, and the chief of police passed judgment on soft bread and the yeast that produced it. They ordained that *pain mollet* could be made, but only with certifiably pure brewer's yeast, since some strains of barm were indeed contaminated.

The issue of contamination lingered for a long time. Even in nineteenth-century America, people distrusted yeast-risen bread, so other types of fermentation were widely used. The available yeast came in the form of barm, or "slop" yeast from potato water. In 1865, Charles Fleischmann, a visiting Austrian with brewing experience, related the poor quality of American bread to its poor fermentation. He began producing a premium yeast, and by 1876, was demonstrating how it made bread rise in a model bakery at the Philadelphia World's Fair. Fleischmann's yeast came in airtight wrappings so it could be readily shipped, making it popular with commercial bakers. That ensured Fleischmann's success (it was yeast-fortune heir Raoul Fleischmann who bankrolled the first issue of the *New Yorker* magazine in 1925) and helped tilt American eating preferences toward bakery bread.

Whereas in baking, alcohol is an unused by-product of fermentation, in brewing, the goal is to produce an alcoholic beverage from fermented grain. This multistage process begins with mash, a mixture of malt (barley specially prepared for brewing) and water. Sometimes plain sugar is added, which can be directly fermented by the yeast; or cereal adjuncts such as wheat or corn, which break down into fermentable sugar. The result is boiled with hops; these dried cone-shaped flowers come from the hop vine, which is related to the marijuana plant cannabis. They add the characteristic bitter flavor and help preserve the beer. After the liquid cools, yeast is added and fermentation ensues. Further processing yields a beverage containing 3 to 7 percent alcohol as well as carbon dioxide produced during fermentation.

There is a huge range in beers. Variations in starting materials, flavoring agents, and processing produce lager, ale, stout, and other types. Each type contains the elements of a foam: bubbles from residual carbon dioxide, proteins that act as surfactants, and polysaccharides, another type of molecule that contributes to foaminess. Polysaccharides are examples of carbohydrates, which as sugars and starches form much of the food that animals and humans eat, and as cellulose make up the supporting structure in plants. A polysaccharide is a polymer, a large molecule made from thousands of smaller units, in this case, units of sugar. Along with resin-containing compounds from the hops, polysaccharides make beer viscous so it drains slowly out of any foam that forms, making the foam last longer.

The particular mix of molecules in a given beer and its effectiveness in raising a head depend on a variety of factors. Beers made with cereal adjuncts, for instance, contain proteins that are more likely to link together and enhance the foam. But whatever the proportions of the molecules, the result is that a foamy head appears when beer is poured, and lasts for some time.

The head is so integral to beer that the beverage would lose much of its appeal without it. (The brewers of stout even add nitrogen gas to enhance the head.) Studies show that beer drinkers consciously or unconsciously notice everything about the foam, such as the size of its bubbles, and whether it adheres to the glass (called *lacing* or *cling*) as the level of beer goes down, which is considered desirable. But the average drinker may not know that foam affects the taste of beer, because it traps some of the compounds that define the aroma of beer. If these compounds are prevented from entering your mouth and reaching the smell receptors in your nose, the flavor of the beer you're drinking is diminished, because smell is an essential component of taste. Therefore the head on a beer can seriously affect your perception of its flavor.

As anyone who has ever poured a beer knows, the head can be controlled: It depends on the height, speed, and angle of pouring, and the shape of the glass into which the beer is poured. Brewers often suggest that their beer be enjoyed in a glass shaped to give the biggest head. But while that carries visual appeal, it might not be the right choice for the best flavor, as shown in a test carried out in 1995. Fifteen professional tasters sampled a cross section of the 400 different beers made in Belgium, which they tasted in four different shapes of glasses. The experts found that traditional big-head cylindrical shapes were best for some types of beers but not for others. Beers with a somewhat fruity taste, for instance, tasted best in gobletlike glasses. Since the head on a beer is also sensitive to any hint of oil or fat, which diminishes the foam, beer glasses should be well washed. They also must be free of any trace of soap, which also reduces the head. Even fat from other foods on the drinker's lips can be a problem.

Brewers put considerable effort into controlling the quality and consistency of the head, to maintain customer appeal, and to ensure that their product looks and tastes the same wherever it is made and served. The Dutch brewer Heineken expects its beer foam to last just five minutes after pouring. A discrepancy of less than half a minute in either direction is enough to cause a newly brewed batch to be rejected. Such changes can come from small variations in the brewing process, which might increase the level of fats or damage the surfactant proteins. The need to maintain foam uniformity has led brewers to fund what might be called beer foam science, which also receives support from champagne makers. Researchers have developed sophisticated ways to study the head on a beer; they measure its height with infrared light, and sample it with a glass fiber that collects light scattered from the bubbles to determine their size—all designed to monitor brewing so it provides the head that makes a happy beer drinker.

GOLDEN *CREMA* AND
WHITE FOAMY MILK

What the head is to the beer lover, the foam called *crema* is to the espresso aficionado. One large European supplier of espresso coffee, the Illycaffè company of Trieste, calls the beverage a "concentrated foamy elixir," which refers to the golden-brown *crema* that signifies a well-made cup of the stuff. Espresso is made by quickly forcing hot water under high pressure through finely ground roasted coffee compacted into a dense mass. (The word *espresso* means "pressed out"; it also refers to the "express" speed with which it is made, and to the fact that it is made expressly for the customer.) The *crema* is produced as the water is pushed through and past the close-packed grains of coffee, although how that happens is not fully understood. And it has taken considerable development to reach the point where the process consistently produces the *crema* that is a sign of quality, and does it quickly enough to attract a paying public.

Originally the water was forced through the coffee by steam pressure. But this powerful force, strong enough to drive trains and defeat John Henry, had to be carefully watched. Each cup was a major production that required delicate control by the machine tender, the *barista*, to keep the high-temperature steam from burning the coffee and making it bitter. Eventually steam-driven machines were invented that could make espresso in large amounts. One model, shown at the Paris Exhibition of 1855 after twelve years of development, produced 2,000 cups an hour; but quality was still lacking, since the result remained bitter.

Slowly, inventors found other ways to produce the necessary high pressure without the harmful effects of steam. Francesco Illy, founder of Illycaffè, tried compressed air in 1935; ten years later, that was replaced by a mechanical piston driven by a lever manipulated by the *barista*; and in 1950 the espresso machine took its modern form, where hot (but not boiling) water is forced through the coffee by an electric pump. This process rapidly produces a cup of rich coffee with *crema* and without bitterness.

To the experienced espresso drinker, the appearance of a substantial *crema* shows that all is well with the coffee and its brewing. (The traditional test is whether the foam can support a teaspoonful of sugar.) *Crema* is important beyond its pleasing look and value as an indicator, because like the head on a beer, it affects flavor. In a recent analysis at the University of Aveiro in Portugal, food chemist Manuel Coimbra, his student Fernando Nunes, and their colleagues brewed espresso from coffee grown in Brazil and in Uganda. They subjected the results to a battery of tests that show espresso to be a rich brew of sugars, fats, proteins, and carbohydrates, including polysaccharides. According to their analysis, the proteins are the surfactants that make the *crema*, and the polysaccharides help by increasing the viscosity of the coffee so that it drains slowly from the foam. The *crema* slows down the release of volatile compounds that carry aroma. That extends the pleasure of the drink by delivering its flavor in what Coimbra calls a "dosed" fashion, meaning it is released over time much as a steady stream of medication emerges from a time-release capsule.

The researchers also showed how to make a foamier and more attractive cup of espresso. They found that heavily roasted coffee beans produced *crema* of a rich, golden-brown color in the greatest amounts. However, the *crema* lasted longest when the beans had received only a medium roasting. That's bad news for those who want their *crema* both plentiful and durable; the degree of roasting that increases one diminishes the other. In that sense, espresso is forever a compromise; its beans can be processed to meet one preference or the other, but not both at the same time.

Crema is not the only complex foam found in a coffee bar. There is also the hot foamed milk that is added to espresso to make a *latte* or a cappuccino. (A *latte* differs from a cappuccino in the proportion of milk to coffee, and in how much of the milk is frothed rather than just heated. *Latte* means "milk" in Italian, and the name cappuccino describes the light-brown color of the drink, resembling that of the habits worn by the Capuchin order of friars.) Unlike the making of espresso, this foaming is still done by steam. A pipe delivers steam deep inside a pitcher of milk, where it produces multitudes of bubbles; these bubbles survive as a foam because milk contains proteins that act as surfactants.

Milk, which is mostly water, looks white and opaque for the same reason foam does: It is full of inclusions, small globules of fat, that scatter light. It also contains several kinds of proteins (as well as sugar, salts, and vitamins) made of amino acids held together by internal bonds. Heating milk under the *barista*'s steam wand denatures the proteins; that is, breaks the bonds so that a spherical protein, for instance, unravels into long coil-shaped molecules, like a ball of yarn clawed apart by a cat. The coils bond with each other in new ways that do not re-create the original molecules, but form a network resembling a surrealistic jungle gym which bolsters the walls between bubbles. The result is a froth that persists even after removal from the steam.

This explains why foamed milk is hard to refoam once it has collapsed; denaturing is not wholly reversible, and each time milk is foamed, its proteins lose some capacity to strengthen foam. There are other complexities in the frothing of milk. The quality of the foam depends on the brand and type of milk (skim, 2 percent, or whole) and its temperature. Most espresso drinks sold in the United States use milk, which has expanded its market as a beverage for adults, and so the dairy and the coffee industries have looked into the best ways to foam it. According to cappuccino mavens from both industries, the best foams come from ice-cold skim milk. Warm skim milk is less frothy than the cold variety, and whole milk at room temperature or warmer hardly foams at all.

However, even with the best choice, cold skim milk, studies show that foamability is affected by how the milk is processed in the dairy. That explains why foamability varies from one brand of milk to another, but one puzzle remains: If milk makes good cappuccino foam, why is it difficult to whip it into a self-sustaining froth?

WHIPPED CREAM

Although vigorous whipping can make milk froth, it never becomes as upstanding as whipped cream, the substantial foam that results when air is beaten into heavy cream (with a little added sugar and vanilla for flavor). According to Harold McGee in his classic work *On Food and Cooking: The Science and Lore of the Kitchen,* this is a case where the fats are more important than the proteins. Cream is 20 percent fat, and these globules stick together to strengthen the bubble walls beyond what the proteins alone can do. The globules also make the cream so viscous that it drains from the foam extremely slowly.

Even with an electric mixer, it still takes a few minutes to whip a batch of cream. Those without the time or inclination can simply press the valve on an aerosol can from the supermarket and watch freshly whipped cream erupt on demand. The aerosol container, representing an ultimate "foam in a can" technology, was once called a bug bomb. This unappetizing name arose when the technique was developed in the early 1940s to deliver insecticide as an aerosol, or a fine vaporized mist. Nowadays the technology delivers all sorts of foams, from whipped cream to shaving cream. In each, gas is mixed with the product to be foamed and held under pressure in a strong metal can. When the valve is opened, the gas expands furiously to create mounds of foam. (Early versions of the valve were prone to clogging. A reliable plastic version was introduced in 1953 by the inventor Robert Abplanalap.)

Americans love instant whipped cream. The makers of the

Reddi Wip brand estimate that 4 million pounds of their product are consumed during November through December, especially on Thanksgiving pumpkin pie. Aerosol whipped creams come in dairy and nondairy versions, both propelled from the can by nitrous oxide (also called laughing gas because it induces mild hysteria when used as an anesthetic). The dairy type contains familiar elements—cream, other milk products, sugar, and corn syrup—as well as an additive or two; but the nondairy type pushes the technology of edible foam to its limits. The cream is replaced by palm oil or other vegetable oil, water, and an imposing list of additives. Depending on the brand, they might include polysorbate 60, sorbitan monostearate, or sodium stearoyl-2-lactylate, which keep the oil and water blended; xanthan gum, which adds texture to the product; and lecithin, a soybean derivative that protects against rancidity. Within this array of compounds, the secret to fluffy whipped cream is that the vegetable oil has a higher fat content than cream, giving a substantial and long-lived foam.

EFFERVESCENT REFRESHMENT

While the rich fat-based foaminess of whipped cream lasts a long time if the cream is chilled, a contrasting kind of foam, thin and short-lived, appears when a carbonated soft drink is poured, especially if the liquid is warm; then the pressurized carbon dioxide gas in the beverage raises a foam that dies so quickly you can see and hear the breaking bubbles.

Soft drinks have not always been carbonated. Pedestrians in seventeenth-century Paris, for instance, could buy lemonade from vendors who sold it from tanks strapped to their backs. Artificially carbonated drinks—now a worldwide industry with annual sales over $50 billion in the United States alone—were created for a more compelling reason than refreshment. They were made in imitation of naturally effervescent waters that poured from certain springs, waters believed to have therapeutic effects. In fact, although the mineral content of spring waters may have value, effervescence per se has no proven medicinal properties, other than the ability to settle nausea.

Nevertheless, people were willing to pay for natural bubbliness in waters such as Perrier, which emerges from a spring near the French city of Nîmes and was first bottled and sold in 1863. Once the bubbliness of spring water was recognized as coming from dissolved carbon dioxide, eminent scientists looked for ways to artificially put the gas into water, creating a carbonated beverage.

The first to succeed was Joseph Priestley, the eighteenth-century British scientist who identified oxygen as the active part of the atmosphere (and also discovered the nitrous oxide that pushes whipped cream out of an aerosol can). Priestley made carbonated water by using beer, or at least the carbon dioxide produced by its brewing. He visited a brewery in the city of Leeds, and holding two containers over the surface of the mash as it fermented, poured water back and forth until the liquid was charged with CO_2. In 1772, he developed equipment to carbonate water more conveniently. At about the same time in France, Antoine Lavoisier, a founder of modern chemistry, developed a similar apparatus.

Entrepreneurs eagerly adopted this new technology. One of the first was the Swiss jeweler Jacob Schweppe. He made and sold carbonated mineral water in Geneva, and later in London, providing two different formulations for medicinal purposes. To prevent the gas from escaping, his product came in corked bottles with rounded bottoms; such a "drunken" container could not stand upright, guaranteeing that its cork would remain damp and swollen so as to provide a tight seal.

Apart from any perceived medical virtues, people drank sparkling water because they enjoyed its fizz and its slightly sour flavor, which comes because carbon dioxide in water yields highly diluted carbonic acid. There is also evidence that the fizzing action and perhaps the acidity affect the taste buds so they are more sensitive to food flavors. The popularity of bubbly water spread rapidly; by 1851, the Schweppes firm was selling well over 2 million bottles a year. A natural next step for other entrepreneurs was to add flavoring, such as lemon in the 1830s. And the first cola soft drink, based on caffeine-bearing nuts from the tropical cola tree, came along in 1886, when John Pemberton invented Coca-Cola in Atlanta, Georgia.

Today, the production of carbonated soda involves massive quantities of CO_2. At Coca-Cola, the world's biggest beverage company, carbonation is not carried out at one central location but rather at regional bottling plants. First the basic Coca-Cola syrup (whose recipe is a well-guarded secret) is blended with water. Then the mixture is chilled to a few degrees above freezing, which makes it easier for it to take up CO_2 and reduces the tendency to foam. The gas is added in stainless steel tanks filled with CO_2, after the liquid has been vaporized in diffusers, which makes it more permeable to the gas. Different brands of soda, and different product lines by the same manufacturer, have different degrees of carbonation, which is controlled by varying the pressure in the tank. The typical value for Coca-Cola is sixty pounds per square inch, twice the pressure in an automobile tire and much higher than atmospheric pressure—a far cry from Joseph Priestley's groundbreaking but low-tech carbonation over fermenting beer mash.

At the opposite end of the luxury spectrum from soda is champagne, a unique sparkling wine that uses the carbon dioxide created during fermentation. Champagne was developed, it is said, in the old French province of Champagne around 1700 by a monk, Dom Pierre Pérignon. His special contribution was a particular blend of red and white grapes. He kept his effervescent wine in bottles made of especially strong English glass and closed with cork, which sealed tightly enough to hold in the gas. By the mid-eighteenth century, the firm of Moët and Chandon was making the wine commercially, and others soon joined in.

The special blend of grapes for champagne still comes from that region in the northeast of France, with the best grapes grown in vineyards along the Marne River and near the city of Reims. The first step in making champagne is the same as for any wine: The grapes are fermented in vats to produce alcohol. But there is a special second step, in which more yeast and sugar are added, to induce further fermentation. In modern champagne production, the second fermentation may be done in stainless steel vats. In the traditional method, however, it occurs in sealed bottles and, after several months, produces enough carbon dioxide to generate considerable pressure inside the bottles. The wine then ages for up to five years, which results in a sediment of dead yeast. To remove the yeast, a process called *dégorgement* is employed, which is labor-intensive for bottle-fermented champagne. Every day, each bottle is shaken, rotated, and tilted a bit farther off the vertical until it is upside down with the impurities settled onto the bottom of the cork. Then the cork is removed, and the internal pressure blows the sediment out.

Dégorgement illustrates the considerable power of the gas trapped in champagne. Its internal pressure is about seventy pounds per square inch, five times the downward push of the Earth's atmosphere at sea level. That is enough to propel a popping cork at a speed that can be dangerous. The pressure has also been a problem in very large bottles of champagne, such as the "balthazar" size, which holds twelve liters (about three gallons). These big glass containers were once handblown but proved prone to explosive breakage, and sturdier glass was introduced in the mid-1980s. However, champagne makers still

contain the gas with the same type of cork used by Dom
Pérignon, which comes from the bark of a tree prevalent in
Spain and Portugal, although now the stopper is held down by
a wire cage to keep it from blowing out. (Efforts are under way
to replace natural cork with synthetic plastic for some types of
wine. That would be a shame for champagne, for the biological
cells in cork look distinctly foamlike. Whatever advantages
plastic stoppers might offer, they lack this pleasing image of
foam held back by foam.)

That contained pressure, of course, also causes the "pop,"
the froth, and champagne's festive bubbling. The best champagne glasses are the tall narrow shapes called flutes; with their
small cross section, they give the deepest foams and hold bubbles the longest. More famous, perhaps, are the shallow
stemmed goblets, three inches across by two deep, that symbolize New Year's Eve. According to legend, they are modeled
on Marie Antoinette's breasts; but unfortunately their design
guarantees quick dissipation of the bubbles.

Studies of champagne foam have been surprisingly limited,
although proteins are known to play a leading role. Only recently have researchers at the Oenology Laboratory in Reims
correlated the properties of the foam with those of the base
wine, that is, the wine made by the first fermentation before the
second one adds the sparkle. They injected carbon dioxide into
base wine to make a controlled foam. Its height and stability indeed depended on proteins in the wine; and on polysaccharides
that thicken the wine so it drains slowly from the foam. The researchers also found an unexpected subtlety: The presence of
iron in the base wine contributes to the foam. The reason is
that iron bonds with proteins to make a stronger surfactant
than the proteins alone. Another metal, copper, plays a similar
role in egg whites, as we'll see in the next section.

Other researchers have examined foam in sparkling wines,
which contain carbon dioxide but are not made by the cham-

pagne process, and obtained informative but somewhat discouraging results. A team at the University of Barcelona looked at forty-eight varieties of the sparkling Spanish wines called cavas. They found that the height and durability of the foam are affected by the type of grape and the blending and aging of the wines. As is true for espresso, where the *crema* depends on how much the coffee is roasted, this knowledge gives us some control over the foam; but as is also true for espresso, it is not possible to optimize all variables simultaneously. Wine aged for nine months gave the tallest foams, whereas wine aged for eighteen months gave the most stable ones.

LIGHT EATING

If champagne is the lightest French beverage, meringues and soufflés are the lightest French foods. Both use egg whites, which nature has developed as the nearly perfect agent for strong and fluffy foams. Egg white is water that contains dissolved solids, mostly proteins, whereas egg yolk has a large proportion of fat and other constituents. A meringue is a mixture of egg whites and sugar thoroughly beaten together with air, then baked until it is dry and turns brown on top. The structure that makes the meringue stiff and upstanding comes from certain proteins in the egg whites; these proteins become denatured as vigorous beating mixes them with air, and they recombine into a delicate network. This traps air that the beating has forced into the egg whites, making the basic foamy structure. The heat of baking denatures other proteins into coil-shaped elements that coagulate to strengthen the structure (if you have ever seen egg white curdle as it emerges from an egg that has cracked under boiling water, you have watched those tiny bedsprings link to each other), and the heat also expands the trapped air to make the meringue puff up.

Over centuries of making meringues, chefs have learned that not even a drop of egg yolk can be allowed into the egg whites, or the volume of the beaten foam is drastically reduced. The culprit is the fat in the yolk. Fats and oils reduce surface tension, which is why oil calms a troubled sea—lower surface tension means that water cannot rise so high into waves. Similarly, egg whites tainted by fat cannot rise high into a fluffy meringue. Also, the fat molecules interpose themselves between proteins that would otherwise combine into a foamy network.

Another bit of lore, dating back at least to the late eighteenth century, is that a meringue whipped in a copper bowl rather than a china one is creamier and less likely to be spoiled by overwhipping. Harold McGee carried out experiments to confirm this, and explains the effect. The minute amount of copper scraped off the inside of the bowl as the egg whites are beaten provides enough atoms to bind with the proteins. Like the iron-protein combinations that enhance champagne foam, these copper-protein complexes are more stable than protein alone, giving a better meringue. McGee also attempted to enhance beaten egg whites with iron, and also zinc, but so far only copper—unfortunately, the most expensive metal of the three—provides the right kind of complex.

A soufflé (the name comes from the French verb *souffler*, meaning "to blow or to puff") is more complex than a meringue. In addition to egg whites, it contains flour, milk, and butter; often egg yolks are included, as are cheese, chocolate, or other additions. The egg whites are beaten to make a froth, the other ingredients are added, and the mixture is baked in the oven, where it rises. The construction of a soufflé represents the most fragile artistry of all the edible foams; Irma S. Rombauer and Marion Rombauer Becker, the authors of the *Joy of Cooking*, write that a well-made soufflé "reminds us of dandelion seed puffs just before they blow." That perfect structure cannot be achieved unless the egg whites are beaten just so, and it is also dependent on another tricky operation: As Julia Child puts it in *Mastering the Art of French Cooking*, the beaten whites must be "incorporated gently and delicately" into the other ingredients to avoid breaking the bubbles in the foam; but they must be incorporated rapidly or fluffiness also suffers, as my own soufflé-making efforts have taught me.

Even vastly experienced master chefs know the difficulties of making a flawless soufflé. John Doherty, for twenty years executive chef at New York's Waldorf-Astoria Hotel, says the recipe must be extremely well balanced ("Everything in a soufflé has a reason," he says), and adds his personal secret, a well-buttered pan to facilitate rising. Laurent Gras, formerly *chef de cuisine* at noted restaurants in Monte Carlo and Paris and now at the Waldorf's Peacock Alley, notes that only the freshest ingredients will do. The age of the eggs matters, he says, and adds that even an accomplished chef must run through a given soufflé recipe twenty or thirty times before adding it to his or her repertoire.

Assuming everything is done perfectly, a soufflé puffs up as much as three times its original size in the oven. Part of that ballooning comes from expanding air, but most comes from the vapor the heat produces from liquid water. Coagulation of the egg, and probably other reactions involving the gluten in the flour, strengthen the foamy structure; they do not, however, make it rigid and long-lasting, since even the best-made soufflé collapses soon after removal from the oven and must be eaten immediately. The reason this happens is not fully understood, but that rapid fall has become a metaphor for failure or collapse in any human realm. It is also the bane of soufflés made for restaurant patrons; as executive chef Doherty says, "You need a waiter standing next to the oven door" to speed the dish to the diner quickly enough to be enjoyed.

It is questionable whether it will someday be possible to make nondeflating soufflés, but technology can help make soufflés more reliably. When a soufflé bakes, its internal temperature rises, then holds steady, and finally climbs rapidly to about 100 degrees Celsius (212 degrees Fahrenheit) as the last of the water steams through the top. At that point the soufflé has completely risen and should be removed from the oven. This critical moment can be properly registered with a device called a thermocouple—a pair of thin wires inserted into the heart of the soufflé, which sense the temperature and convert it into a voltage that is measured outside the oven. Such a computerized soufflé might seem right only for the galley of the starship *Enterprise*; but even master chefs use thermometers to check their cooking, so perhaps a cybersoufflé would not violate the canons of *grande cuisine*.

Certainly kitchen technology does not yet show serious signs of turning soufflés into fast food. The closest anyone has come is to make soufflé mixtures that can be refrigerated or frozen, later to be warmed or defrosted and baked in the oven. These prefabricated foams may not rise as high as those made on the spot and whisked into the oven, but some do well even under microwave heating, which also engenders novel dishes made with edible foam. One such recipe is for a Frozen Florida, which resembles an inside-out Baked Alaska.

In a Baked Alaska, a block of hard ice cream is placed on a layer of cake and coated with uncooked meringue. The meringue is quickly browned in an oven, emerging warm to give a wonderful contrast with the still-cold ice cream. In a Frozen Florida, the meringue is inside, made into a hollow case filled with a mixture of jam, sugar, and brandy. The case is coated with chocolate icing, and the whole affair is frozen. Just before serving, the Florida is placed in a microwave oven. The microwaves heat the central alcohol-and-sugar mixture, but the insulating properties of the foamy meringue prevent the heat from reaching the icing. The result is an unusual dessert with a hot center surrounded by cold chocolate, the inverse of cold ice cream within a warm coating that is a Baked Alaska.

ENJOYING THE TASTE
OF KITCHEN TECHNOLOGY

Edible foams, from bread and beer to a Frozen Florida, are special in two ways: through their pleasing lightness and texture, and through their role as among the most artificial forms of food. Whereas fruit, fish, and wild grain are edible with little or no processing, foamy foods, from beer to nondairy whipped topping, require considerable human intervention. The brewing, baking, and other processing that produces them may seem unexciting compared, say, to modern digital technology. But we take kitchen technology for granted precisely because it has been part of human culture for so long. In its own way, it is highly evolved, and has had lasting value beyond how we eat and drink. The nineteenth-century study of fermentation, for instance, can fairly be called one of the roots of biochemistry, a central discipline of the life sciences.

Edible foams also illustrate the limits of fundamental science. Take milk foam, which depends on the behavior of proteins. We have identified the exact proteins in milk, and know how they link to each other, and how they change under heat, which explains how they maintain a foam. But when it comes to actually making a cappuccino, nothing so profound comes into play; rather, the right or wrong brand of milk may be the determining factor. That suggests a rhetorical question: "Can everyday foams, like milk foam, ever be fully understood and controlled?" The answer seems to be, "Understood, yes; fully controlled, perhaps not." And for edible foams where the scientific study is less advanced, or which contain a wickedly complex combination of compounds as in espresso, we cannot even claim full understanding.

Many scientific characteristics of foam enter into the food and drink that it animates, from the lifetime of its bubbles to its thermal behavior. But when you come down to it, it is the sheer fact of mixing gas with liquid or solid that makes edible foam so pleasing. The mixture provides an airy elegance of presentation that enhances the enjoyment of dining. More important, the mixture gives the unique texture that makes a taste of foamy food or drink a satisfying sensual experience, the elusive quality that food chemists and gourmet tasters call "mouthfeel." Fortunately, our empirical kitchen knowledge of foam is enough to give us wonderful soufflés, beer, and champagne.

Although it would be a stretch to claim that kitchen knowledge carries over directly into the uses of nonedible foams, these do depend on similar techniques and qualities. The processing that turns a semiliquid mixture of flour and water into solid bread is like the procedures that turn a liquid solution of certain molecules into a solid foamed plastic. The insulating properties of the meringue in a Frozen Florida are inherent in the nature of foam; they are the reason you can hold a hot cappuccino in your hand within a plastic coffee cup. The fact that whipped cream stands up on a cake but flows obediently under a spatula is the same property that makes shaving cream useful. In the next chapter we look at these and other desirable uses of foam, as well as the less happy cases where foam diminishes rather than enhances how we live.

PRACTICAL FOAM

CORK, AEROGEL, AND SHAVING CREAM

In the classic 1967 film *The Graduate,* Dustin Hoffman's bewildered just-out-of-college character Benjamin Braddock endures a patronizing lecture by a middle-aged businessman who delivers a one-word message—*plastics*—as his recipe for success. That was not what Benjamin wanted to hear, but the businessman was right, in the sense that plastic has become steadily more important since it went commercial in the 1930s. In 1977, the United States for the first time produced as many cubic feet of plastic as of steel, and in 1979, the same could be said for the whole world. Since then there has been no turning back. Plastic has come to infiltrate every part of our lives, from grocery bags to computer cases to automobile parts, with a global usage of nearly 300 billion pounds in 1997.

For all its weight in the world's economy, plastic has yet to completely escape certain chintzy and even sleazy undertones. They linger from the days when the material, less robust and varied than it is now, was thought of as a cheap substitute for something better. It would be a mistake, however, to forget that foamed plastic, along with other kinds of foam, has real technological value (and in the duality that technology always seems to bring, carries environmental costs as well).

From transportation to medicine to telecommunications, technology relies on natural and synthetic materials adapted to human use. Over the centuries, we have learned to mine, grow, make, and process whole families of materials, each with its own characteristics. Metals are hard and strong, take an edge for cutting, and efficiently carry electricity; glasses and ceramics insulate against electricity and withstand extremely high temperatures; organic compounds, made of carbon, hydrogen, and often nitrogen or oxygen, can be tailored into useful substances from drugs to dyes.

Among these materials, solid and liquid foams bring their own special properties. They can be both strong and lightweight. They can insulate against heat, cold, and sound. They can be made hard or soft, firm or resilient, and more—or less—like a liquid or solid. They can be held in a compact form and quickly expanded to a large volume; their bubbles or pockets can provide immense internal storage. In solid foams, that inner space comes in two different geometries, each with its own advantages: closed-cell, where the gas cells are individual units like so many soap bubbles; and open-cell, where the voids are interconnected.

PUMICE, SPONGE, AND CORK

Even before foams could be made to specification, people took advantage of the foamlike characteristics of such natural materials as pumice, sponge, and cork. Sponge and cork do not come from an actual foaming process, but pumice is a true solid foam. Made the same way as some foamed plastics, pumice is formed as gas trapped in hot liquid lava expands into bubbles that become fixed as the lava cools and solidifies. Volcanic eruptions are often accompanied by this rocky froth. For example, when Mount Etna erupted in 79 C.E., Pompeii was buried partly in pumice.

Some forms of pumice contain so many gas-filled cavities that they are less dense than water and can float. (Lightness characterizes artificial solid foams as well, extended as they are by gas of negligible weight.) The cavities make pumice an excellent abrasive, as it has been used at least since Roman times. When a piece of the stuff is broken, the edges of the cells form a hard jagged surface that is ideal to abrade or polish material. Pumice is still used in heavy-duty soap and abrasive pads that scrub away stains or dead skin. It defines contemporary fashion in stonewashed jeans, which are buffed into pleasing softness as they are tumbled in washing machines along with pieces of pumice.

The sea creatures called sponges have vast internal storage capacity, another feature of foamy geometry (in this case, the open-cell type). For centuries, divers have gathered sponges from the Mediterranean Sea; in classical times, sponges—or rather, the flexible skeleton that remains after a sponge dies— were used as brushes, and as drinking cups by Greek and Roman soldiers. The ability to hold quantities of liquid comes from the internal design of these immobile creatures, once thought to be plants. Now they are classified as primitive ani-

Natural sponge (left). Cellular structure of sponge (right).

mals with a skeleton defining an intricate structure of many connected channels.

These channels are essential for the sponge to feed by ingesting nutrients from water it continually pumps through itself. That is not done by muscular action, which is how the human heart pushes blood through the body, but by tiny whiplike appendages called flagella, only micrometers long. These move large amounts of water because there are so many of them. The internal area of a typical sponge is sixty times greater than its external surface. This large inner surface supports enough beating flagella to sweep a fresh supply of water through the entire creature every few seconds. When a Roman centurion quenched his thirst from a sponge, he was drinking water held within that large internal area. And when he squeezed the sponge to take up water or expel it, he was using its flexible open-cell geometry.

Artificial sponges emulate these properties with a flexible matrix full of interconnected cavities, but need not copy the exact arrangement of channels. In any foamy structure, no matter what the shape of its cells or whether they are separated or connected, the same geometric truth applies: Break a space into small volumes, and you increase its internal surface.

To see why, imagine you are an architect charged with converting a large cubical loft into an avant-garde showplace, where art will be displayed on the floor and ceiling as well as the walls. How do you accommodate the greatest amount of art? If each of the six inner surfaces—floor, ceiling, and four walls—covers 1,000 square feet, leaving the loft as a hollow cube gives 6,000 square feet of display. But if you split the loft into two galleries by building a horizontal divider halfway up the walls, you add a floor and ceiling, and the display area jumps to 8,000 square feet. And if you now run a vertical divider from top to bottom, creating four galleries, that adds two more walls for a total of 10,000 square feet. The more you subdivide a given volume, the greater its internal area, and that fact is useful in many applications of foamy materials.

Cork illustrates other qualities that artificial foams emulate. Like sponges, cork is closely associated with the Mediterranean region. It is the bark of the cork oak tree (the word *cork* comes from the Latin *quercus*, meaning "oak"), which grows mostly in that area. The tree was once widespread in Spain and North Africa; after the Spanish Civil War and the colonial troubles in Algeria, however, Portugal became the center of the cork industry. It now produces over a billion wine corks a year, not a new use of the material; cork stoppers have been found in Greek wine jars over 2,000 years old.

Examine a thin slice of cork through a microscope, and you can see why it seals so well. It has a closed-cell structure, with

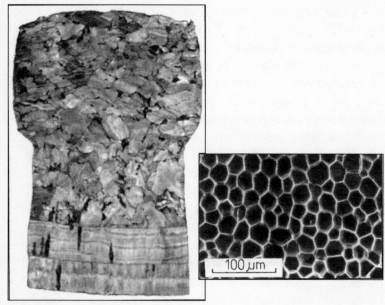

Cross section of a champagne cork (left). Cellular structure of cork (right).

cells filled with air and separated from each other by walls made of a waxy substance. This makes cork resilient: Press a wine cork between your fingers, and it deforms slightly; release it, and the cork springs back. The cells and the trapped air compress to a certain point as force is applied, and then exert outward pressure that tries to return the cork to its original volume. As you force a cork into a wine bottle, it compresses, and then expands as far as possible, completely sealing the neck of the bottle.

Its resilience also makes cork a good choice for shoe soles. From the platform shoes worn by the 1940s film star Carmen Miranda, to the antichic 1960s look of Birkenstock sandals, cork provides springiness. In addition, the trapped air makes cork so light that it floats high; a slab a foot thick rides on water with only slightly over two inches immersed. The waxy cell walls are waterproof, so cork does not become water-logged. For these reasons, cork has long been used in life preservers.

The cells also make cork an excellent soundproofing material. Sound waves reach the ears by traveling through air and any other media they encounter. You might think sound travels best through a soft medium like air rather than a hard one like steel, but the opposite is true. Pluck a taut guitar string, and the energy you impart moves along smartly, soon setting the whole string vibrating to produce a tone. Pluck a slack guitar string, however, and no sound emerges, because the energy of plucking is quickly absorbed as it travels between the string's loosely connected elements. Similarly, the loosely linked molecules of air in cork, and the soft walls of the cells, soak up sound more than they convey it. The result, as Marcel Proust appreciated when he wrote *Remembrance of Things Past* in his cork-lined study, is that cork effectively suppresses the noisy distractions of the world.

Cork has another useful property that reappears in synthetic foams: It is a poor transmitter of heat. There are only three ways to send heat from place to place: radiation, conduction, and convection. The Sun's warmth reaches us through 93 million miles of nothingness by means of radiation, another way of saying that a hot body generates electromagnetic waves which carry energy through space. The metal handle of a saucepan can burn the hand because the heat applied to the pan is efficiently conducted along the handle. And the area in front of a heating vent is pleasantly warm because of the physical motion or convection of hot air (which also occurs in liquids). Cork does not transmit radiated heat because it is opaque. It is a poor conductor because of its air-filled cells. And the air trapped in those cells cannot support convective motion. As a result, cork is an effective thermal insulator.

"JUST TWO WORDS: *FOAMED PLASTICS*"

The useful qualities in pumice, sponge, and cork point to synthetic solid foams that have similar qualities but with improvements, such as the ability to be easily shaped for diverse applications. Foamed plastic, also called cellular or expanded plastic, meets many of these needs.

Foamed plastic is made from ordinary solid plastic, which comes in bewildering variety, from polyethylene to polyurethane to polycarbonate, and more. The common link, and a clue to the basic nature of plastic, is the prefix *poly*. A plastic is a polymer, which means it contains giant molecules made from many smaller molecular units called monomers. The monomers are generally organic in nature, and the name for each plastic indicates its special monomer. For the simplest plastic, polyethylene, the monomer is the ethylene group, two linked carbon atoms with two hydrogen atoms hanging off each carbon. For polyurethane and polycarbonate, the monomers are the more complex urethane and carbonate units respectively.

Plastic is made in the process called polymerization, where the monomers link to each other in varied geometries, each containing thousands of monomers: linear, with monomers arranged like beads on a string; branched, where the central string sprouts side branches; and network, which contains closed loops as in a fishing net. In some cases, the long main strands can be cross-linked, meaning additional molecular bridges are formed among them. This strengthens the polymer, much as if you were to add diagonal bracing to the vertical and horizontal bars of a child's jungle gym, enhancing its rigidity and stability.

By selecting the starting monomers, their linking geometry, and the degree of cross-linking, chemists can make plastics soft or hard, clear or opaque, optimized for strength or flexibility or lightness or smoothness. In each case, a batch of monomers is mixed in liquid form. The molecules link to each other, and then the liquid hardens, either naturally or through further processing, to form solid plastic.

If air or other gas is added while the plastic is liquid, the result is foamed plastic. Pockets of gas are trapped as the plastic hardens, yielding a solid foam. The gas can be introduced in any of several ways that uncannily resemble the preparation of edible foams. In some foamed plastics, air is beaten or stirred into the monomers, much as whipped cream is made by beating air into cream. In others, heat is applied to produce gas, like the final stage of baking a soufflé, except that the gas is usually nitrogen (the harmless element forming nearly 80 percent of our atmosphere), not water vapor. In still others, the gas comes from a foaming agent (also called a blowing agent) added to the mix—just as yeast added to flour and water produces carbon dioxide—or from the monomers themselves, like self-rising bread.

Foaming makes plastic more versatile, because the ratio of gas to plastic can be widely varied. A cubic foot of solid polystyrene weighs sixty-two pounds (by coincidence, the same as a cubic foot of water). The foamed version can be made with so little gas that the weight drops to sixty pounds; or with so much that it drops to only a pound. The choice of closed or open cells gives further control over the properties of the foamed plastic. Open-cell types tend to be soft, since the gas they contain does not resist applied pressure but simply moves to another part of the internal maze; hence open-cell foamed polyurethane makes comfortable cushions. Closed-cell types are only slightly flexible because the gas resists compression beyond a certain point, and can even be rigid if the plastic matrix is strong enough.

Plastic foams reproduce the good qualities of cork with improvements: They insulate against heat and cold, and provide comfortable sitting or walking, but do not rot and are not attacked by living pests; they can be made strong and rigid if desired; and whereas cork comes only as sheets cut from trees, plastic foams can be made in any shape. This gives enormous flexibility in applications. Instead of lining a room with sheets of cork, for instance, noise is now stopped right at the ears by tight plugs of flexible polyurethane foam. The resilient foam polyvinyl chloride (PVC) can be molded to form the soles of sandals, where it provides a tougher surface than cork.

Some high-profile applications dramatically illustrate the usefulness of foamed plastic, although with the occasional drawback. The material has been essential in the U.S. space program, where it insulates the bullet-shaped fuel tank mounted externally on NASA space shuttles. Half the length of a football field, this enormous container holds 1.5 million pounds of rocket fuel and the substances that oxidize or burn the fuel, extremely cold liquid hydrogen and liquid oxygen. Both must be protected from heat generated by air friction as the craft streaks upward at 25,000 miles an hour. Heat is again an issue when the empty tank plows back down through the atmosphere after being jettisoned from the rising shuttle. Then the concern is that heat not build up too rapidly, which could destroy the tank and scatter debris in populated areas rather than over empty ocean.

This would seem an exemplary use of foamed plastic, since it provides essential insulation without adding much weight, a critical concern for the shuttle. But it had unexpected consequences in June 1995, when the shuttle *Discovery* was being prepared for launch from the Kennedy Space Center. The craft's fuel tank had been coated with rust-red urethane foam. With the shuttle pointed straight up, the tank apparently resembled an extra-big tree to a male yellow-shafted northern flicker woodpecker, an aggressive species of bird standing twelve or fourteen inches high. To impress potential mates, or maybe out of sheer nearsightedness, the bird furiously pecked up to 200 holes in the foam, up to four inches across. That compromised the thermal protection and introduced other problems. The launch was delayed for weeks while the damage was repaired, at a cost of $2 million or more.

Foamed plastic provides excellent heat insulation even in ordinary take-out coffee cups. (These are often called Styrofoam cups, but "Styrofoam" is actually Dow Chemical's brand name for its particular form of polystyrene foam, used by florists and in decorative applications. Take-out cups and other fast-food products are made from a different kind of foamed polystyrene.) It is not just that you can hold hot coffee in your hand and feel only warmth; the material protects against temperatures far above or below room temperature.

Consider the peculiar substance called liquid nitrogen, a cousin of the liquid oxygen and hydrogen used in the space shuttle, whose temperature is only 77 degrees above absolute degree (more than 200 degrees Celsius or 390 degrees Fahrenheit below room temperature). With widespread industrial and scientific uses, liquid nitrogen is stored in insulated containers because it instantly boils away if it touches anything at normal temperature, which by comparison is white-hot. However, an ordinary take-out cup insulates well enough to hold liquid nitrogen for some time. I regularly use these throwaway items for laboratory experiments at low temperatures.

At the other extreme, a polystyrene coffee cup is the salvation of at least one racing car driver who must cope with high temperatures in the cramped confines of his car. The 1996 Rookie of the Year, Johnny Benson, found that his feet ended up directly over his car's exhaust system, which operates at hundreds of degrees. He insulates his feet from the hot floorboards by cutting an ordinary coffee cup in half lengthwise, and fitting a half-cup to the heel of each foot inside his shoe, allowing him to work accelerator, clutch, and brake in comfort.

Your fingernail can dent a take-out cup, and that crazed NASA woodpecker easily carved holes in urethane foam.

Foamed plastics, however, can be made harder and stronger than that, while being formed into any desired shape, for instance, by being foamed within a mold. The molding process can be manipulated to yield a structural kind of foam, with a tough solid skin covering the more fragile internal cells. Such methods have led to new ways to make furniture, since urethane foam can be made strong enough to sit on, and handsome enough to please. At New York's Museum of Modern Art, a 1995 exhibit called "Mutant Materials" featured several chairs made of the foam, coated with wood film for a decorative look.

Other novel uses of foamed plastic appeared in Italy after World War II, where designers set new directions in consumer items. They expressed their aesthetic in plastic, among other materials, and the result was innovation that often rebelled against tradition. A recent sale at Christie's in London brought together several foamed examples of the period and the attitude. One "seating system" from 1966 consisted of five polyurethane cushions of varied form that could be used for sitting or simply piled up for aesthetic effect. (A more radical seating system, although it did not use foam, consisted of a patch of bright green plastic grass.) Another designer, Gaetano Pesce, came up with a "squished" chair that could be achieved only with foamed plastic. This polyurethane item came crammed into a small flat package. When the package was opened, the (presumably) delighted purchaser could watch the furniture swell up to full size.

These applications shade off into smaller *objets* that combine usefulness with a pop art–like foamed kitschiness. One prime example is a coatrack exhibited recently at the Atlanta International Museum. Created around 1971 by the Italian designers Guido Drocco and Franco Mello, the rack has functional arms on which to drape clothing; but arranged in a fanciful shape inspired by those icons of the American West, Arizona-style cacti.

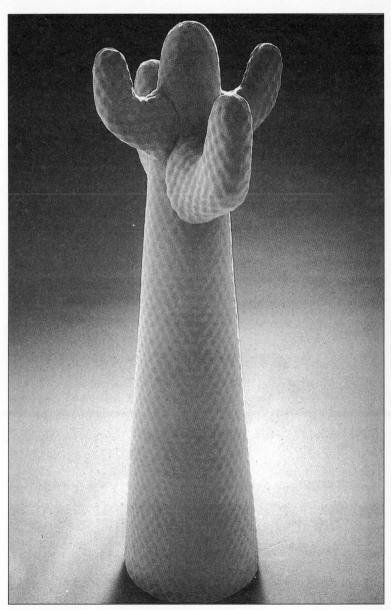

Cactus coatrack made of foamed plastic.

The coatrack's stubby, nicely rounded limbs are made from structural polyurethane foam, coated with tan lacquer for protection and appearance. Although the effect is silly—the rack resembles a real cactus, huge, green, and covered with spikes, about as much as the kid's purple TV dinosaur Barney resembles a real T. rex—it is charming.

The kitschiness of foamed plastic can have a more serious meaning in some artistic contexts, such as when its frothiness is put into sharp contrast with a deeper tradition. The New York–based artist Brian Tolle carves foamed plastic to replicate structures that resonate in American history. His 1996 installation *Overmounted Interior* was a Colonial-style interior scene complete with a faux fireplace and fake wooden ceiling beams; he has also created fences of sculpted foam that seem to be made of irregular pieces of fieldstone, and which would look perfectly at home in historic New England.

PACKING AND LANDFILLS

It is a tribute to the versatility of foamed plastic that its uses extend from the aesthetic to the cartoonish to the pragmatic. And nothing is more pragmatic than the protective material you encounter when you unpack a new stereo system, which comes perfectly nestled into foamed plastic blocks; or a delicate vase, surrounded by inch-long bits of foamed plastic. Each of these plastic peanuts, made of closed-cell foamed polystyrene and called loose-fill in the trade, is a tiny shock absorber. Under impact, it compresses against the resistance of the gas trapped in the cells, dissipating the force before it does any harm. In one test of cushioning power, an egg survived unbroken when it was dropped from twenty feet onto a pad of foamed polystyrene less than an inch thick.

According to John Mellott, president of Storopack, a major

maker of packing peanuts, they got that name because early models tended to uncoil from an "S" shape into a lumpy form. The idea of free-flow protective packaging may have been inspired by popcorn, once used by the U.S. government as a homey kind of loose-fill. Now, fueled by the exponential growth in shipping generated as Americans increasingly order from catalogs or on-line, 45 million pounds of peanuts are made per year in the United States. That's an astonishing amount, enough to cover the borough of Manhattan three inches deep. You might think this plastic blizzard, which has produced well over a trillion peanuts, contains only identical clones; but these little artifacts come in engaging variety. They appear in colors from white to pink, and in specific shapes; the letters *S* and *E*, the figure *8*, a contact lens-like form, and others.

Shape and color pretty much summarize a given peanut. Says Orwill Hawkins, president of Inter-Pac, another big producer, "I can pick up a peanut and tell you who made it." Each manufacturer turns out its product in forms meant to absorb shock and reduce settling during shipping, forms which also become company logos in plastic. Storopack uses the "S" shape and others. Free-Flow, a third large maker, favors an "8" shape, which begins when polystyrene is extruded through a die into long hollow tubes. These are crimped into a figure-8 cross section, sliced, and expanded in a steam bath into Free-Flow's own special peanuts.

The colors also mean something. Appropriately, white peanuts come from virgin polystyrene, or recycled pure white packaging. Off-white or tannish gray ones contain other recycled material such as coffee cups. Pink ones have been treated to reduce static electricity, which can damage electronic equipment. Green peanuts from some manufacturers mean 100 percent recycled content. From others, however, they mean only that green dye has been added to give an implicit suggestion of environmental goodness.

Environmental impact is an issue, because for all its usefulness and pop cultural weight, foamed plastic (along with other plastics) fares poorly in the environment—or rather, fares all too well. Each packing peanut, coffee cup, and food container thrown into the trash and taken to a garbage landfill stays there forever. Most man-made materials deteriorate under the long-term effects of sun, rain, and wind. Solid plastic takes many decades. Foamed plastic, however, endures: it does not disintegrate into simpler parts even under acid rain. Not ever.

People became aware of the undesirable longevity of foamed plastic around 1989, when McDonald's clamshell food containers began piling up in profusion. No one has yet come up with a way to make food containers or plastic peanuts disappear more quickly. The industry's recycling of polystyrene helps, and so does the "Peanut Hotline," which consumers can call to find sites that accept piles of peanuts for reuse. But many people are disturbed by the idea of an unnatural material that hangs around forever. Free-Flow's Virginia Lyle, who manages communications and environmental affairs for the company, regularly hears from those who consider plastic peanuts "the spawn of the Devil," as she puts it.

Spawn the peanuts may be, but they are not all that prevalent in our national garbage heap, which has been analyzed in the long-term Garbage Project at the University of Arizona. After picking through 125 tons of garbage, the researchers reported in 1992 that foamed polystyrene, including peanuts and food containers, takes up 1 percent or less of the total landfill space in the United States. When the researchers added in every other type of plastic, they found up to 24 percent, depending on the type of fill; but concluded that the large amount of paper may be an even bigger problem, since it does not deteriorate rapidly either.

AEROGEL AND FOAMED METAL

Although foamed polystyrene is light, it falls far short of the world's lightest solids, the foamy materials called aerogels. Consisting of 98 percent air or more, they have been called the soufflés of materials science. A photograph of an aerogel shows a ghostly image of something barely there, aptly described as "frozen smoke." Aerogels are chock-full of desirable properties. An ounce contains the area of several football fields, which could be used for such applications as storing rocket fuel and holding electrical energy by storing electrons. The material provides excellent sound insulation, and an inch of aerogel blocks the transmission of heat as well as ten double panes of ordinary glass. Since slabs of aerogel are also transparent or translucent, they could be ideal as thermal windows or skylights in cold climates, allowing sunlight to heat a building while blocking heat loss to the outdoors.

With these wondrous characteristics, you might ask why you're not seeing aerogels all around you. The reason is that although they were developed in the 1930s, only now are researchers beginning to make them easily, cheaply, and safely. An aerogel begins as a mixture of solid and liquid called a gel, like Jell-O or the substance of a jellyfish. Close examination shows that a gel is a three-dimensional mesh outlining interconnected cells filled with liquid. It resembles a liquid-filled sponge with tiny cells, only ten to twenty nanometers across. (A nanometer is a billionth of a meter, the size of ten hydrogen atoms laid side by side.)

To turn gel into aerogel, you remove the liquid but keep the solid network, now filled with air. That is tricky, since the liquid partly supports the delicate mesh. If it is not removed with care, it pulls the network as it departs and the whole unlikely structure collapses. The first aerogel was produced by an intricate process that included making a gel from silicon dioxide (SiO_2, also called silica), the main constituent of sand and of ordinary glass, and hydrochloric acid. Then the acid was replaced with water, and in turn the water was replaced with alcohol. The result was put into an autoclave, a strong closed container, where it was heated to 240 degrees Celsius (460 degrees Fahrenheit) under high pressure, equivalent to that under 250 feet of water. The autoclave treatment drove out the alcohol in a way that left the mesh intact.

This tedious method of manufacture stretched over days and took its toll, for autoclaves can be dangerous. In 1984, an enormous unit was set up at the University of Lund in Sweden to make quantities of aerogel for its main scientific application, the detection of elementary particles. The seal on the autoclave failed, and the result was an explosion that destroyed the installation.

But now aerogel can be made more safely and quickly, using silicon dioxide, metallic and organic compounds, or carbon, each giving different properties. Scientists at the Lawrence Berkeley National Laboratory have decreased the chances of catastrophe by reducing the necessary temperature and pressure, and others have made aerogels without an autoclave. With signs that production is becoming routine, companies have considered using aerogel as thermal insulation in refrigerators, acoustic insulation in automobiles, and both kinds of insulation for aircraft, where light weight has its greatest value. Even without commercial use, aerogel is invaluable for science, with an essential role in NASA's Stardust mission to gather material from the early solar system, and in NASA's missions to Mars, as will be described in chapter 7.

An aerogel might be defined as a solid foam made by adding something to nothing, that is, using the merest wisps of solid to divide a volume of air into small cells. Adding nothing to something also pays off, when that means introducing gas into solid metal. The basic idea is expressed in metal girders made lighter by having holes drilled into them, without seriously diminishing their strength. It reaches full fruition in foamed metals, which are ordinary metals like aluminum shot through with tiny openings.

Such a porous metal has the expected advantages of a foam: It is light, with large internal storage and good thermal and acoustical properties. What's remarkable is that subtracting metal adds an unexpected benefit. The melting point of solid aluminum is much lower than that of steel (although still high), a factor that must be considered in applications such as warship design. But foamed aluminum resists high temperature better than the solid material, presumably because the network of air-filled spaces slows down the transfer of heat.

One way to make foamed aluminum (or tin, zinc, and other metals) is to combine it with a foaming agent that releases gas when heated. The mixture is brought to a high temperature, the metal softens, and the agent makes hydrogen bubbles that remain after everything has cooled, much as happens in foamed plastic. This can be done within a mold, producing a foamed piece shaped to specification. In another method, air jets play through molten aluminum, producing a foamy top layer that is siphoned off onto a moving belt to make sheets of material.

The ERG Materials and Aerospace Company, of Oakland, California, does it differently. For thirty years, ERG has used a proprietary process to produce an open-cell structure in aluminum, copper, and other materials. The cells are bigger than in aerogel, but still many thousands appear in a cubic inch of the aluminum foam the company calls Duocel. (The name came from the mistaken belief that the basic cell is shaped like a dodecahedron, a solid with twelve faces.) The material is fascinating to look at and handle; the delicate three-dimensional tracery of the metal is evident, and adds an appealing surface texture.

Duocel was originally used in hush-hush Cold War applications. There had been times, according to Bryan Leyda, ERG's engineering manager, when representatives of various federal agencies were "sitting right here at this table dividing up our production capabilities." One secret use was in reconnaissance satellites that carried sensitive infrared sensors to detect missile launches and other military activity. The sensors operated at temperatures near absolute zero, provided by on-board reservoirs of liquid helium. With its large internal surface and good thermal characteristics, Duocel was ideal to trap the cold helium within its cells. That kept the liquid from sloshing heavily, and perhaps causing damage, no matter how the satellite gyrated in space, while carrying the cooling power of the liquid helium exactly where it was needed.

With the Cold War over, Leyda is now considering varied uses for Duocel. One possibility is dual-purpose airplane wings, where the cells hold fuel. Another is prosthetic devices. According to Leyda, it is not just that foamed metal is lighter than the metal now used in artificial limbs. Mechanically speaking, Duocel is also more like certain bones in the body with an internal cellular structure (as discussed in the next chapter). This means that the joint where metal joins bone and flesh is less likely to be stressed to the point of failure.

LIQUID FOAMS

As we have seen, solid foams display a tremendous range of properties, with hardness, for instance, varying from the sharp-edged grittiness of pumice to the comfortable cushiness of polyurethane. None of them, however, can flow. Only liquid foams flow, spread, and cling, which suits them for a whole second universe of applications. Like solid foams, with their natural antecedents of pumice, cork, and sponge, liquid foams have natural forerunners such as the froth on a wave. But it is hard to relate their useful characteristics to this ephemeral natural example.

Chemical engineer Andrew Kraynik, a leading researcher in the area, knows the difficulties better than most. Kraynik has been thinking about liquid foam for the twenty-odd years he has worked at Sandia National Laboratories near Albuquerque, New Mexico, operated for the U.S. Department of Energy. Starting with a straightforward interest in how foam flows, he found that subject impossible to grasp without knowing all about the static mechanical properties of foam, which depend in turn on its innate geometric structure. Kraynik finds that fully connecting this bedrock knowledge to the dynamics of foam and therefore to applied liquid foams requires answering enough questions to occupy him for decades to come.

That hasn't prevented Kraynik from talking about flowing foam at the Gillette Company and working with Dow Chemical to extend the links between understanding and application. And it hasn't prevented chemists and inventors from developing liquid foams that do everything from shave beards to muffle explosions.

SHAVING CREAM

With typical human ingenuity, people keep finding unexpected applications for shaving cream. It has been recommended as a congenial sculptural medium for children, and as a substance that can be used to fill in highly eroded inscriptions on old gravestones, making them legible without harming them.

Shaving cream eerily resembles nondairy aerosol whipped cream: It is propelled by gas out of a can to make mounds of fluffy white stuff, and it contains a variety of odd-sounding compounds, even sharing one, polysorbate 60, with the whipped cream. Some ingredients do more than make shaving easier; they smell good, soothe the face, and offer other enticing if ill-defined benefits. One shaving cream on the market is laced with rosemary, sage, avocado oil, and Ginseng root. Anything that gives a ginseng buzz while seeming good enough to eat could be an excellent product to keep around; but maybe not for those with toddlers, who tend to think shaving cream looks deliciously edible.

Amid these blandishments, it can be hard to remember the simple, plain-vanilla function of shaving cream, which is to keep the beard wet. According to the Gillette Company (whose shaving cream has the distinction of being the medium in which diffusing-wave spectroscopy was first tested at Exxon), beard hair is as hard to cut as copper wire of the same thickness. The remedy is to soak the beard in water for a couple of minutes, which makes the hair expand and soften, and easier to slice. Just splashing on water will not do, because it soon drains away or evaporates. Foam, however, is pliable enough to spread over the face, and stiff enough to adhere, keeping its water in contact with the beard as long as necessary. That makes the hair 70 percent easier to cut, according to Gillette. Foam also makes shaving more pleasant because it helps the razor blade glide over the skin.

LIFESAVING FOAMS

The ability to bring and hold water where it is needed is what makes liquid foams so effective in fighting fires. Although fire-fighting foam carries relatively little water, containing up to 95 percent air, it brings the liquid into close contact with the fire while also smothering it. In gasoline fires, where plain water would sink beneath the burning fuel and spread it, foam rides atop the blazing gasoline to snuff out the flames. Firefighters need not worry about suffocating those trapped in confined spaces such as mines and the holds of ships because people can breathe through the air-rich foam, and the limited amount of water also minimizes collateral damage to property.

Like any kind of foam, fire-fighting foams need surfactants. Some use protein, which works just as well as it does in meringues whipped up from egg whites. This doesn't mean that every fire extinguisher contains a dozen eggs or requires expensive protein from a filet mignon. Cheap and inelegant proteins from chicken feathers, ground-up cattle hooves, or pigs' blood are fine, as are proteins from plants. Although plant proteins don't fully meet people's nutritional needs unless certain food combinations (like beans and rice) are used, the protein in even a single vegetable like the soybean (also known as soya) is a perfectly adequate surfactant.

Two important and lifesaving soya-based foams were developed by the pioneering American chemist Percy Lavon Julian. Julian, grandson of a former slave, was born in 1899 and, against considerable odds, earned degrees in chemistry at De-Pauw and Harvard Universities and at the University of Vienna. He made his reputation in 1935 by performing the first chemical synthesis of physostigmine, used to treat glaucoma. Despite his achievements, discrimination kept Julian from suitable academic or industrial positions until 1936, when he became director of chemical development at the Glidden Company in Chicago—the first African-American in charge of a substantial industrial laboratory.

Julian's initial act at Glidden, a maker of paints and varnishes, was to replace milk proteins in Glidden products with cheaper soya proteins, enhancing the company's profitability. Then he turned to soya-based foams. In Germany, chemists had learned to process soybean oil to yield sterols, the basis of the synthetic male and female hormones progesterone and testosterone then used to prevent miscarriages and treat cancer. Although superior to earlier methods using cattle brains, the

Percy Lavon Julian

process was slow and costly, until Julian turned the solid oil into a porous foam. That allowed the solvents that extracted the sterols to interact with the oil over a large area, speeding up production and lowering costs. Julian went on to turn soya proteins into surfactants in the fire-fighting agent Aero-Foam, widely used by the U.S. Navy in World War II and called *bean soup* by sailors in honor of its vegetable origins. Aero-Foam was especially valuable aboard aircraft carriers, with their huge stocks of highly flammable aviation fuel, and tankers.

Julian's own life suggests another use for foam. Despite numerous scientific and academic honors, when Julian moved to Oak Park, then an all-white suburb of Chicago, his home was first burned and then bombed. Now we all live in a world where a bomb obliterates an airliner and its passengers, as happened over Lockerbie, Scotland; where a van-load of explosives demolishes a building and its occupants, as in Oklahoma City. In times when terrorism is an increasing threat, researchers at Sandia National Laboratories are studying how to use liquid foam to suppress explosions, an antiterrorist application that can conceivably save lives.

When an explosive shock wave encounters a barrier of liquid foam, much of its force is expended in blasting the foam into droplets of water and then converting the liquid into vapor. In one test at Sandia, five pounds of the military explosive C4 (one of the *plastiques* favored by terrorists, and at five pounds enough to do major structural damage) was detonated under a mound of foam like that used for fire fighting. The foam absorbed 90 percent of the destructive energy developed by the explosive. In a similar vein, a quick-hardening rigid polyurethane foam is being tested at Sandia for use in nullifying mines on land or in water by buffering soldiers and equipment against their explosive force, or to lay down a safe ribbon for vehicles to travel.

Another antiterrorist foam prevents unauthorized entry into critical sites like nuclear facilities by encapsulating intruders so they cannot see or hear. And as part of a project to invent non-lethal military and police weaponry, Sandia scientists have come up with "sticky foam." Sprayed from a nozzle to wrap around a person's limbs and torso, it sets and clings so stubbornly to the skin and to itself that movement becomes nearly impossible. It produces no lasting ill effects, although the foam could turn deadly if it covered the nose and mouth. The U.S. Marines who went to Somalia under United Nations auspices in 1992 carried sticky-foam guns, which they used when the force withdrew in 1995.

Sandia researchers have also invented a foam designed to neutralize chemical and biological agents that terrorists might use, providing protection for sensitive installations like airports. The foam carries chemicals that disrupt even virulently toxic military nerve gas, the most horrific weapon of this type.

(One such gas, sarin, killed twelve people when it was released in the Tokyo subway system in 1995, in an attack linked to the Aum Supreme Truth cult.) Although the foam could not save anyone whose skin had been touched by even a droplet of nerve gas, its rapid expansion would effectively confine any outbreak of the gas. The foam also deactivates poisonous mustard gas, first used in World War I, and seems equally effective against bacteria and viruses that can be spread to cause disease.

The rapid expansion of a liquid foam is also the basis of an innovative medical application: the stopping of bleeding in cases of severe trauma by extending the blood's own natural ability to clot. When a blood vessel is torn, an enzyme called thrombin combines with the blood protein fibrinogen to make fibrin, a sticky, meshlike substance that forms a barrier to bleeding. This can be insufficient, however, if the veins and arteries are badly mangled. In such cases, sheer loss of blood is a serious outcome; it is said to be responsible for half of all deaths due to gunshot, whether in the commission of a crime or on the battlefield.

To deal with injuries of this magnitude, researchers at the American Red Cross have developed a foam called fibrin sealant that can be sprayed into a wound even under battlefield conditions. As it expands, the foam carries thrombin and fibrinogen throughout the damaged region. Much as occurs when separate epoxy and epoxy hardener are mixed to make a tough glue for home repairs, the thrombin and fibrinogen combine to initiate the clotting reaction and control bleeding wherever it is occurring. The clots form within seconds, increasing chances for survival until full medical help is available.

Another property of liquid foams, the ability to reduce friction between two surfaces, has been crucial for many a jetliner that has had to make an emergency landing, approaching the ground at upwards of a hundred miles an hour. Foam is readily available at airports, which keep huge fire engines on alert, ready

Test aircraft being stopped by foamcrete.

to spray quantities onto a fire or a runway. In one recent incident, an Air France MD80 made an emergency landing at Antalya, Turkey, after an engine failed. With its landing gear also inoperative, the aircraft slid on its belly to a safe stop on a foam-covered runway, without injuring its 173 passengers and crew.

Foam at airports also provides the air-age equivalent of a railroad bumper, the sturdy structure that marks the end of the line in a train terminal. That is foamcrete, a mixture of cement, water, and foaming agents. Placed at the end of a runway, it is soft enough so the wheels of a moving jetliner plow into it, and tenacious enough to drag the aircraft to a quick stop. In tests, a Boeing 727 was brought to a halt from sixty miles per hour in 400 feet of foamcrete—only three times the distance needed to brake an automobile traveling at the same speed. Foamcrete is projected to be used at several airports, including New York's

La Guardia, where in the last nine years three aircraft have sped off the ends of runways and into Flushing Bay.

DRILLING FOR OIL

Lubrication from liquid foam helps in the big business of drilling for oil, where a rotating bit, like the steel bit on a home electric drill, cuts deeply into rock until it reaches the oil reservoir. This operation usually proceeds more or less vertically (wells have been bored nearly six miles down) but must sometimes be done horizontally. The latter subjects the rotating equipment to considerable friction, which can be minimized by filling the well with foam.

Vertical oil wells are also kept full of liquid as drilling proceeds, circulating between the surface and the bottom to remove the rock cuttings. In oil fields where the drill might encounter high pressure from gas trapped with the oil—which can "blow out" the hole, causing a gusher that must be capped—the fluid is a dense slurry of water and clay called drilling mud, heavy enough to prevent gushers. But in depleted fields, where pressure is low and gushers unlikely, a lighter liquid is used so pressure from the formation can actually push oil into the well.

One formulation used in this "underbalanced drilling" is basically sparkling water, except charged with nitrogen rather than carbon dioxide. But foam is better, according to tests carried out by the Canadian oil company Tesco Incorporated at the Jean Marie formation, which covers much of northeastern British Columbia. The foam could be adjusted to maintain underbalanced conditions and was far better than water/nitrogen at cleaning out rock cuttings. That substantially improved drilling rates, an economic bottom line in the business: The drill penetrated twenty-seven feet per hour on average with foam, over twice as fast as with water/nitrogen.

UNINTENDED FOAM

Along with these useful applications, there is the unintended foam that sometimes appears. Motor oil can develop a foamy nature that reduces its ability to lubricate, many industrial processes develop unwanted foams, and strange foams appear in sewage. One impressive type of sewage foam was first seen in the 1960s in Milwaukee, and in England since. This brown foam, called chocolate mousse by sewer experts, forms atop the settlement tanks that are part of sewer systems. It can support the weight of a cat and expands so quickly that it nearly climbs out of the tank, or, as one British observer put it, "It can go walkabout." The foam seems to come from soil bacteria that wash into sewers and grow into long filaments that trap air to produce foam. It may be tied to the increasing consumption of polyunsaturated fats as people eat vegetable oils rather than animal fats. This is thought to alter the makeup of human waste so that it is more appealing to these particular bacteria.

Foam is also an indicator of polluted rivers and streams. Pure water contains no surfactants that can sustain a foam, so its appearance means something else is present (unless the water is agitated and mixes with air, as in white water). That something may be perfectly natural, like the dissolved salts and organic compounds that sustain sea froth; but it may be man-made, as happened in the Pigeon River. The Pigeon, which starts in the Blue Ridge country of North Carolina and continues into Tennessee, is one of the older and more bitterly contested examples of waterway pollution. Its contaminated condition began receiving national attention in the 1980s, and later Al Gore, himself a Tennessee native, became involved.

The river has been used since 1908 by the Champion International Corporation of Canton, North Carolina, to make paper and carry away the by-products of the process. Residents along the river had long learned to live with discolored water and a lack of aquatic life. But when the pollution began to be thoroughly examined, a truly serious health hazard emerged: The few fish in the water contained the chemical 2,3,7,8-TCDD, one of the highly toxic family of compounds called dioxin that can cause cancer and damage the reproductive and immune systems. The dioxin entered the Pigeon from the chlorine that Champion used to bleach its paper.

Almost every account of the Pigeon River at its worst (much of the pollution has now been cleaned up) notes its striking foaminess. A 1989 story in *U.S. News and World Report* describes how Champion "spewed out a dark, foamy effluent"; the Tennessee Environmental Council says, "The 1990's still found a river . . . where pockets of yellow foam would form in the eddies"; a woman born near the Pigeon and who raised her family there recalls the river "had foot-high white foam floating along the top." The foam was said to be so pervasive that barrels were strung across the river to break it up. Along with the discoloration and the smell of the contaminated water, this was evidence that something was terribly wrong.

In Pigeon River and similar cases, the polluting chemicals are the true problem; the foam is only a sign. But it is fair to ask if foam, whether floating on water or residing in a landfill, always carries environmental bad news. The answer is complex, because foam can fit on both sides of the ledger. Some foams act like an oil-absorbing sponge and clean up marine oil spills. If the pollution occurs in air, the noxious gas or smoke can be passed through a foam scrubber, where bubbles trap the offending particles. And it has been suggested that a foamy torrent of sonoluminescentlike bubbles could break pollutants such as dry-cleaning fluid into harmless components.

Ironically, one form of foam can even hide the environmental effects of another. This application was developed by "Foamman," the *nom-de-Internet* of Paul Kittle, a chemist. Kittle consults on the uses of liquid foam, which he calls "a solution in search of a problem," the comment made about lasers when they were first invented. His Web pages list dozens of applications from varied sources. Many have environmental implications, such as the possibility that foam could extinguish the twenty-year old Percy Coal Mine fire in Pennsylvania, and the use of an acidic foam to destroy asbestos installed in buildings by simply spraying it on.

In the late 1980s, Kittle worked out a way to cover garbage landfills with foam. A significant portion of a landfill is occupied by plain dirt, which according to EPA guidelines must be piled six inches deep every night to cover that day's trash. Kittle came up with an environmentally benign shaving cream–like foam that would adhere even to steep slopes and would not blow away. The foam stopped rats and bugs, and prevented odors from rising. But unlike dirt, it dissipated after thirty-six hours, no longer taking up room when it was no longer needed under newer trash. For this reason, says Kittle, using his foam could save up to 15 percent of landfill space. He has since moved on to other uses of foam, but his invention leaves behind an image to relish: the huge tracked vehicle Kittle designed, patiently spreading liquid foam to cover acres of garbage made partly of indestructible foamed plastic peanuts, coffee cups, and McDonald's clamshells.

DANCING IN CHAMPAGNE

Foam can hardly seem more serious and depressing than when it is in or on a heap of trash. Important as these environmental issues are, we should not lose sight of the lighthearted uses of foam. In fact, people keep coming up with new ways to enjoy it; for instance, by filling a dance floor with foam. Beginning, it is said, on the Spanish island of Ibiza in the late 1980s or early 1990s, the fad soon spread across the Atlantic. By the mid-1990s, clubs across the United States were running Freaky Foam or Foam Frenzy nights, when dancers would be immersed in four to eight feet of foam conjured up from baby shampoo or from a solution said to be used in the movies.

Dancers loved the feel of the foam. One happy foamie said it was like "dancing in a glass of champagne." As a bonus, the bubbles transmitted the bass beat of the dance music so it could be felt as well as heard. For the psychologically minded, dancing in foam represented a strange, and maybe exciting, mixture of the childlike and the sexual. At one club in San Diego, dancers laughed like children while flinging foam at each other, and a club owner in Boston wanted his patrons to bring their rubber duckies. At the latter club, though, something more adult was suggested as the white stuff spurted out of a "foam cannon" and onto the dance floor at regular intervals; everywhere, dancers hidden under the foam could share a private moment in an activity they would not dare perform in the open.

If this particular foamy pleasure is for adults (even if they play at being children), other foams and bubbles are purely for children, as in blowing soap bubbles, a perennial favorite. Or look at any McDonald's with an attached play area, and you're likely to see a bin full of colored plastic balls, which lets kids burrow into a bubbly atmosphere without getting soap in their eyes. Foam for kids reaches higher levels just outside the city of Atlanta, at the million-dollar American Adventures playground. There children get to enjoy both solid and liquid foams: They toss around foam balls and balance them in an air jet, and paint and sculpt with colored foam.

*   *   *

From light uses to serious ones, many properties of technological foam trace back to natural pumice, cork, and sponge, and sea froth; but if we focus only on applications, even the most important, we miss deeper meanings in these forerunners. Pumice and sea froth are clues to the physical processes that made and still shape our planet. (We will examine these processes in chapter 6.)

In the same vein, cork and sponges offer clues to life on Earth. The cellular structure of living things was first seen in cork over 300 years ago by the English scientist Robert Hooke. Sponges may represent a special rung on the evolutionary ladder, a transition between simple single-celled animals and advanced organisms. By exploring the foamlike structure of cork and sponge, and other ways that foam and bubbles appear in living things, we gain insight into life.

five

LIVING FOAM

CELLS, VIRUSES, AND
MEDICINAL BUBBLES

Many of the basics of foam science were established in the nineteenth century, the same period when the science of life was venturing beyond the descriptive study of plants and animals to approach its modern form. By the mid–nineteenth century, the living cell was being recognized as the microscopic basis for life, and in the late 1890s, compelling evidence confirmed that the cell is the smallest unit of living things. A hundred years later, we explore life at several levels of size and complexity, from the molecular scale to the whole living creature, and use this fundamental knowledge to deal with disease and enhance human well-being.

Except at the molecular scale, foam and bubbles enter into all these levels. A single cell displays some characteristics of a bubble, and the collections of cells that form tissues and organs resemble an aggregation of bubbles. Some bodily organs, such as the lungs, have foamlike features that are essential for their functioning, whereas sponges depend wholly on their many internal voids. True foams, bubbles within a liquid, enter into the complicated activities of animals, such as the nest-building practices of certain fish, and into human biomedical practice as well. And sometimes bubbles or foam do biological harm or represent harmful agents, as in the blocking of a blood vessel by a bubble and in the disease of leprosy.

THE CELLULAR MATRIX OF LIFE

Although there are many successful one-celled organisms such as bacteria and amoebae, more highly evolved plants and animals contain hundreds to trillions of cells. These cells are closely linked and often adjoin each other, forming specialized structures such as the muscles and the liver. In 1839, the German biologists Theodor Schwann and Matthias Schleiden recognized that both the individual cell, and the joining of cells into tissues, organs, and organisms, are essential for life. They called cells the "elementary particles" of plants and animals, as atoms are the basis of matter. Schleiden wrote further that "each cell leads a double life: an independent one, pertaining to its own development alone; and another . . . insofar as it has become an integral part of a plant," which is also true for animal cells.

A single cell resembles a fluid-filled bubble, for the cytoplasm it contains—a solution of organic molecules that carry out its life functions—is enclosed within a cell membrane like the skin around a soap bubble. (The cell also generally includes a nucleus containing genetic DNA and other structures.) To form the parts of a complex organism, cells crowd together like bubbles in a foam. The cells making up the leaf of a plant like the hyacinth resemble roughly rectangular tiles and nestle together so as to determine the overall shape of the long narrow leaf. The cells of a blood vessel fit together like tiles as well, in the form of distorted hexagons that are nearly diamond shaped.

This close-packed foamlike geometry gave the first hint that life is cellular, as seen by the seventeenth-century English physicist Robert Hooke. Like Isaac Newton, Hooke worked on such fundamental issues as the motion of celestial bodies and the nature of light. The two scientists crossed intellectual swords, and Newton later avenged himself by a snide reference to Hooke's unprepossessing frame. (There seem to be no portraits of Hooke extant, but a biography written in 1705 describes him as "crooked and low of stature. . . . He went stooping.")

Hooke lacked Newton's analytical genius, but his physical insight and his brilliance as an instrument maker made him an outstanding researcher. Hooke's law is still used to describe how an elastic spring responds to a force, and Hooke was involved in attempts to develop a spring-driven marine chronometer, to solve the old problem of finding a ship's position at sea. Hooke also took important steps toward understanding light and naturally seized on a then-novel optical device, the microscope, which had been invented around 1600. He used a characteristically clever method to make microscopes with especially high magnifications. Hooke melted the end of a thin glass rod and allowed surface tension to pull the molten glass into a small, sharply curved sphere as it solidified. The result was a powerful lens through which Hooke examined details of the natural world.

In 1665, Hooke—barely thirty years old—published his influential *Micrographia*, a compendium of his microscopic observations. The book is stunningly illustrated with Hooke's own beautiful drawings. (Hooke used his artistic talent to help Sir Christopher Wren rebuild London after the Great Fire of 1666, designing Bedlam Hospital.) Most striking are the tipped-in images of a flea and a louse as seen through the microscope, each folding out to over a foot long. At this enormous magnification, they strike an unwary modern reader as unimaginable aliens from a science-fiction movie, somehow inserted into this 300-year-old book.

At the time, such pictures were dazzling insights into the living world. But Hooke made a bigger breakthrough when, taking a sharp penknife, he cut an "exceedingly thin" slice of cork. His microscope showed the cork to be "all perforated and porous, much like a Honey-comb, but that the pores of it were not regular." That porous look came from a swarm of tiny compartments, over a thousand per inch, that Hooke dubbed *cells* from their resemblance to rows of monk's cells, as is evident in his punctilious drawings. Hooke observed similar cells in other plant matter such as carrots, and in the shaft of a bird's feather.

Although Hooke's observations led to the very concept of cell, they were based on the dead bark of a cork tree. By 1683, Antonie van Leeuwenhoek had gazed through his own microscope to see living cells in motion, but all Hooke saw was empty little boxes whose function puzzled him. The English physician William Harvey had already established the circulation of blood, and Hooke thought the cells might circulate the fluids a plant needs to grow. He searched for the necessary openings or valves but found none because they do not exist. Material moves among the cells in an organism as molecules diffuse from cell to cell, that is, work their way between the molecules making up the cell walls. Although Hooke could hardly have discerned that process at the time, in one sense he understood cells extremely well. Drawing on his deep knowledge of mechanics, he cleverly related the lightness and resilience of cork to its cellular makeup, illustrating some of the benefits of a foamlike arrangement for living systems.

But Hooke could not explain the origin of this cellular geometry, which he likened to a "solid or hardened foam" and to a "congeries of very small bubbles." We now know that cells grow into bigger structures in a well-planned manner that follows blueprints contained in an organism's DNA. This growth begins as a fertilized seed or egg divides and subdivides into cell after cell, which develop and link together into complete tissues and organs. Still, Hooke's description had meaning, for the physical processes that determine how cells cluster follow the principles of foam science, as the innovative biologist D'Arcy Wentworth Thompson pointed out long after Hooke's time.

Born in Edinburgh, Scotland, in 1860, Thompson can be seen looking out from his photograph on the Web site maintained by the University of Dundee, where for thirty-three years he held the chair in biology (following which he held the chair in natural history at Saint Andrews University for another thirty-one years). His full beard, piercing gaze, and dark suit with vest make him the very image of a Victorian professor. And he represents a Victorian kind of scholarship rarely seen today; trained in both zoology and classics, Thompson combined his two passions in the translations he made of Aristotle's biological writings and the books he wrote about the birds and fish of Greece.

Thompson also had a bent toward mathematics and physical science, which led to his massive labor of love, the book *On Growth and Form*, published in 1917. Grounded in Thompson's great knowledge of biology, it provides enormous insight into cells and their aggregations, brilliantly arguing that life can be understood through rational scientific principles. The book has become a classic, appearing in a revised edition in 1942, followed by a huge paperback version in 1992, weighing in at 1,100 pages and two and a half pounds of wondrous text and illustration. Not only does Thompson write vividly and lucidly, but the book's drawings and photographs are a fantastic atlas of the shapes that life takes, from cells to butterflies' wings, seashells, and skeletons.

Early in *On Growth and Form,* Thompson expresses his philosophy: "Cell and tissue, shell and bone, leaf and flower, are so many portions of matter, and it is in obedience to the laws of physics that their particles have been moved, moulded and conformed." In chapter 1, we saw that the intricate tessellation of adjoining soap bubbles in a foam is sculpted by the forces acting on it; Thompson sees similar physical sculpting at work in living systems.

Thompson's book appeared just as the details of biological inheritance were being uncovered. Only six years earlier, in 1911, the geneticist Thomas Hunt Morgan had used fruit flies to show that inherited factors are represented by genes at specific sites along a chromosome. But even the second edition of *On Growth and Form* appeared a year before the genetic role of DNA was discovered, and long before James Watson and Francis Crick found DNA's double helix.

Today we know, as Thompson did not, that the design of any living thing is chemically coded in its DNA. Cells arrange themselves according to this imperative, but always subject to physical law; that is, DNA lays down plans that are followed by matter and energy. In the nonliving world, we are used to this combination of abstract template and physical reality. Think of structures made of stone: a wall, an Egyptian pyramid, an arched doorway. Each contains the same kind of material, each is subject to the same laws of gravity, friction, and adhesion; but each follows a different plan and hence takes a different form. In a living organism, the plan is imposed by its DNA.

Without knowing the genetic role of DNA, Thompson nevertheless explained a great deal about biological arrangements through the principles of foam science. He pointed out that individual living cells, like bubbles, are shaped by surface tension. Just as for a soap bubble, the minimizing principle (presented in chapter 1) insists that a cell take the form that minimizes its energy. For unicellular creatures, that means either a sphere or a symmetrical form if there are constraints that prevent a spherical shape. Thompson notes in his book, however, that surface tension is important even for amoebas, which lack a fixed shape. These microscopic unicellular animals move and feed by extending pseudopods in one direction or another. Wherever the creature throws out one of these little tentacles, its surface tension must be reduced, indicating that even this tiny speck of cytoplasm exhibits subtle control over its surface chemistry.

Remarkably, Thompson's insights into these living analogues of bubbles also illuminate the nonliving versions. The irregular shape of a moving amoeba might seem an unmistakable sign of life, qualitatively different from the globular perfection of a soap bubble. But Thompson points out something we have all witnessed, perhaps without taking much note of it: A soap bubble shows ceaseless change. He writes, "The surface of the bubble . . . thins and it thickens, its colours change, currents are set up in it." When we say a bubble is "at rest," notes Thompson, we mean that its changes take place slowly—changes that, like the behavior of amoebas in motion, are connected with variations in surface tension.

These connections go deeper. Both the skin of a soap bubble and the membrane surrounding a cell have surfactantlike molecular arrangements that control surface tension and are also permeable. The permeability is the reason a soap foam coarsens, as molecules of gas diffuse from small, high-pressure bubbles into large, low-pressure ones. In biological systems, permeability allows molecules to diffuse among cells. This intercellular trade carries information, such as the nerve impulses that traverse the neural network, and permits cells to work in concert. As molecules move between muscle cells, for instance,

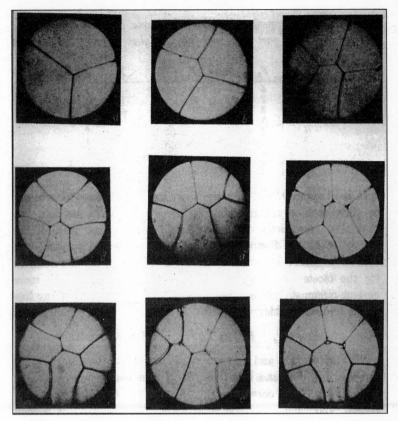

Cell divisions of soap bubbles.

they generate a shared electrical current that signals the cells to contract simultaneously, producing muscular action.

Thompson goes further to analyze how cells abut each other to form living structures. One example is epithelial tissue, where connected cells form a continuous sheet such as the lining of the gastrointestinal system and the epidermal layer of the skin. The forms of the cells are easy to see in these thin tissues, and Thompson shows how their shapes, such as distorted hexagons, conform to Plateau's rules.

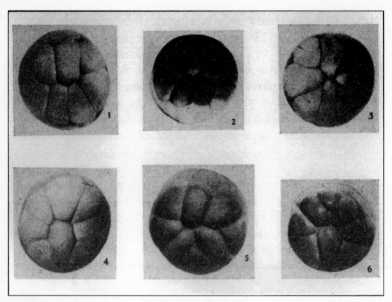

Cell divisions of frogs' eggs.

Most striking of all is Thompson's illustration that a fundamental process of life, the subdivision of one cell into many, follows the same physical principles obeyed by a foam. In a series of compelling photographs, he shows that the arrangement of cells in a frog's egg, as it goes through its first few stages of subdivision, closely resembles that of soap bubbles blown in the two-dimensional environment of a shallow dish—a vivid illustration of how living creatures take on the close-packed cellular arrangements Robert Hooke had seen centuries earlier.

BREATHING THROUGH BUBBLES

These cellular arrangements form biological tissues and organs, many with their own foamlike features. Sponges are defined by their internal void-filled geometry, and in our own bodies, the lungs, certain bones, the kidneys, and the breasts are full of voids or bubbles. These organs draw on the desirable large internal surface areas, flexibility, and lightness of foams.

The bubble-filled geometry of the human lungs helps them take in air, distribute its oxygen so the body can consume food, and expel the resulting carbon dioxide. The lungs contain some 300 million small air spaces called *pulmonary alveoli* (the singular is *alveolus*, Latin for "small hollow"), arranged like bunches of grapes. They are lined with a surfactant, a mixture of fats and proteins that provides lubrication, reducing the muscular effort needed to move the lungs in and out. Like any surfactant, it adjusts surface tension, keeping small alveoli from collapsing into big ones, which could cause unstable breathing. Lung surfactant is critical for the 65,000 premature infants born yearly in the United States, whose systems are insufficiently developed to produce it. Synthetic surfactants now available constitute an important advance in the treatment of premature babies.

It is not obvious why humans benefit from lungs filled with alveoli. They are susceptible to pneumonia, and they hold gas at higher pressure than in a large empty lung, so it takes more work to breathe. But the alveoli offer a decisive advantage: Like the rooms in that three-dimensional art gallery we encountered earlier, they subdivide space. This results in an immense internal surface area approaching 900 square feet, half the size of a suburban ranch house, which is essential for the lungs to exchange enough gas to keep the body going. Oxygen enters the blood as its molecules diffuse through the thin walls of the alveoli and into blood capillaries contained there. This slow process can meet the body's needs only because the huge area allows large amounts of gas to move into and out of the blood with each breath.

The geometry of the lungs varies with animal species. Salamanders have undivided lungs, whereas the lungs of the more highly developed toads, birds, and mammals are subdivided. The degree of subdivision reveals differences in animal metabolism. Some types of frogs have lungs subdivided into spaces ten times bigger than the alveoli in mammals, which influences their respiratory efficiency. One of these frogs typically exchanges gases over fifty square inches of lung area for every cubic inch of air it breathes; a human has fifteen times as much area, over five square feet, devoted to the inflow of oxygen for every cubic inch of air. The higher rate of flow is essential for warm-blooded mammals. It takes energy to maintain that body temperature, and so compared to cold-blooded creatures, mammals must be well stocked with food, and with oxygen to turn the nutrients into usable form.

Cross section of a human femur (left). Cellular structure of a femur (right).

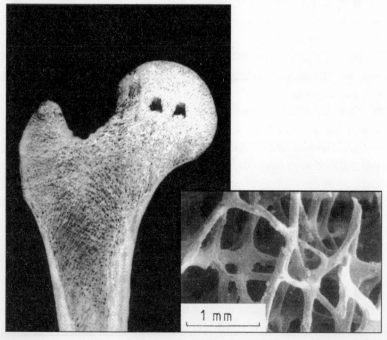

1 mm

One animal in particular, the sponge, depends utterly on a void-filled geometry for its existence. As we discussed earlier, a sponge extracts nutrients from water it pumps through itself by means of the small whiplike flagella that line its internal voids. The ability of the sponge to feed and grow depends on the total amount of water it can move, which is determined by how much internal area it possesses that can support the flagella.

Zoologists organize sponges into three levels of geometric complexity. The ascon type (which means "skin" in Greek) is a single hollow tube, with its internal area no bigger than its external one. That structure limits the number of flagella and hence the feeding capacity, and these types are typically only a millimeter in size. At the next level, the internal structure becomes a set of canals, giving a larger area for feeding. In the most complex type, the canals are replaced by many small chambers connected by conduits and lined with cells that pump water and extract nourishment from it. The enormous area inside a sponge with this structure—typically sixty times its external area, as noted earlier—permits it to feed efficiently. A sponge of this variety may grow a thousand times larger than the ascon type, up to a meter across.

The subdivided space within these animals represents a deeper issue as well. Sponges are ancient creatures. They provide some of the oldest animal fossils, dating back 650 million years. They are more complex than unicellular organisms, since they contain different types of cells; yet more primitive than true multicellular beings, for the cells are not tightly organized into organs like a stomach or liver. In this light, the internal geometry of sponges shows evolution within the species and may tell us something important about how simple creatures become complex.

Nature also uses the fact that a solid foam can be both strong and light in weight. Consider, for example, the construction of cancellous bone, which appears in the human femur, or thighbone, and in the radius and ulna, the two parallel bones in the forearm. Cancellous bone has a smooth, dense outer layer surrounding an arrangement of rods or plates honeycombed with openings, like girders that support the roof of a large free-span building while leaving considerable open space. Cancellous bone, however, adds an ingenious directional element. Its larger elements lie along the directions of the main stresses on the bone, providing strength where needed and saving weight where possible.

PARENTAL FROGS
AND TURNED-ON QUAIL

Although the living systems described so far have foamlike features, none is a true foam. Plants, animals, and humans also use actual gas-filled bubbles in liquid, either singly or clustered into a foam, to sustain and enhance life. Among plants, the freshwater algae called spirogyra, water-silk, or (less appetizingly) pond scum, literally rises and falls on bubbles. During the day, it takes the sun as it floats on the water's surface, buoyed by the bubbles of oxygen that photosynthesis produces. At night, the bubbles diminish and the water-silk sinks, rising again the next day when the sun comes up.

Some insects are even more dependent on bubbles, using them to enter and leave water, and to survive within it. The blackfly (whose biting swarms are said to be capable of killing a chicken) develops to adulthood in flowing water, then uses an air bubble to ascend to the surface and fly off. Other insects turn a bubble into a miniature aqualung. Insects take in oxygen and expel carbon dioxide through their trachea, a system of tubes connected to the outside air through openings called spiracles. Certain predatory water beetles use their trachea underwater. When the beetle dives, it traps a bubble of air under its wings, next to its spiracles. As the beetle consumes the trapped oxygen, the oxygen pressure in the bubble decreases, allowing more of the gas to enter from the surrounding water. The bubble acts like the gills in a fish except that it eventually shrinks. But like scuba divers replenishing their air tanks, some species of beetles trap additional oxygen-filled bubbles from algae.

Foams are prevalent in the complex reproductive and parental behavior of species from insects to birds. The larvae of the hopping insects called froghoppers or spittlebugs are equipped to make spittle, a froth that protects them from predators and keeps them moist. Using a special valve on its abdomen, a larva mixes air bubbles with surfactantlike secretions to create this foam, which is said to become so prevalent in Africa that it drips from tree branches. In another example, the female praying mantis lays up to 200 eggs at a time in a liquid foam, which hardens into a case that armors the eggs against bad weather and determined enemies.

Frogs and toads especially use foam in varied ways to safeguard their eggs and provide for their young. Some, such as certain poisonous South American frogs (whose venom is reputedly used by native tribes to tip poison arrows), carry their eggs with them, stored in backpacks made of foam. Others build floating nests. In one family of frogs found in South America and Australia, the male fertilizes the female's eggs while both float on the surface of the water. The male kicks his legs to stir up a froth of water, air, semen, and newly emitted eggs, forming a floating nest three or four inches across. Its outer surface hardens and becomes meringuelike, keeping the interior sufficiently moist for the eggs to develop. When the tadpoles hatch, they work their way down into the water through the foam, which by this time is starting to collapse.

Other frogs lay their eggs on land, held in a foamy mass that supports the emerging young until rain washes them into a pool. In Zimbabwe, certain tree frogs deposit their eggs into foamy nests that hang over water. The eggs, each a couple of millimeters long, are concentrated near the center of a spherical ball of foam. Researchers speculate that there may be a reason for this arrangement beyond protection. The wet foam at the center of the nest holds enough dissolved oxygen for the 900 or so eggs. But as the outer layer of foam dries and the wet core shrinks, the amount of oxygen decreases. This may force the frog larvae to compete for oxygen, eventually stimulating them to leave the nest and drop into the water beneath—a biological timer that sets the reproductive calendar for the next generation of frogs.

Fish have evolved their own elaborate uses of foam. At least one, the soapfish, a smallish creature found from the Atlantic to the Pacific, uses foam defensively. When agitated, it produces a lather containing a toxic protein that makes the fish inedible. But most fish that use foam do so to guard their eggs. The male Siamese fighting fish constructs a protective froth bubble by bubble. He takes a bubble of air into his mouth, where he coats it with a sticky mucuslike secretion that acts as a surfactant. The result is a sturdy bubble that the fish blows to the surface, repeating the process until a floating nest is formed. Then the male picks up the eggs in his mouth from the bottom, where they sink after fertilization, and blows them into the nest, remaining nearby to watch for intruders.

Foam has reproductive meaning among birds as well, at least in the Japanese quail, a small, plump, brown-beige-tan migratory bird that spends it winters in India and Southeast Asia. In this species, the male produces foam that is transferred to the female along with semen during copulation. The foam seems to increase the chances for successful fertilization, although its exact effects on the sperm are still under study. Elizabeth Adkins-Regan of Cornell University uses the foam produced by this species as a marker to study some significant issues in the reproductive process.

Adkins-Regan describes the reproductive froth as a "stiff white meringuelike foam." It begins as a clear viscous liquid secreted by a specialized gland, which is whipped into froth by rhythmic contractions of a sphincterlike muscle, much as vigorous beating turns egg whites into meringue. The interesting questions center on how the male is stimulated to produce the foam, which happens when he sights a female even if the two are separated by a wire screen. This research may shed light on the nerve path that turns the sight of a female into muscular contractions in the male—a contribution to our understanding of how neural stimulation becomes motor action. Although Adkins-Regan's work focuses on birds, she believes her efforts may contribute to a conceptual framework that will be useful in studying human fertility, with potential benefits for the technology of assisted reproduction.

THERAPEUTIC BUBBLES

Bubbles and foam are already utilized in medical practice. One example occurs in the medical use of ultrasound, sound waves vibrating at rates beyond the upper limit of human hearing, 20,000 hertz (cycles per second). These waves penetrate the body and can give an image of its interior without the harmful effects of X-rays. The imaging is done in much the same way that you shout "Hello" from the rim of a canyon, then listen for the echo to estimate the distance to the opposite canyon face. Similarly, ultrasound waves traveling through the body are reflected when they encounter a change in composition, as at the boundary between two organs. A receiver records these echoes; a computer analyzes them to determine the shape, location, and motion of internal structures; and the result is put into pictorial form for examination by a physician.

Applications of ultrasound imaging include monitoring the fetus in pregnant women, diagnosing the state of the heart and kidneys, and examining blood flow. In many of these cases, the quality of the image can be improved by using bubbles. One example shows promise for the detection and treatment of cancer; it is the noninvasive assessment of tumors for malignancy by ultrasound. Both nonmalignant and malignant growths develop additional blood vessels, which take on organized patterns in the former, but are chaotic and disorganized in malignant tissue. The difference can be clearly seen in ultrasound images where bubbles have been used to enhance pictorial contrast.

In one version of the technique, pioneered in the 1980s in the People's Republic of China, carbon dioxide gas is injected directly into the artery that supplies blood to the suspect tumor, so that CO_2 bubbles travel through the tumor. In another approach, special liquid agents carrying gas-filled bubbles are injected into the veins. The bubbles are small and travel freely without any danger of blocking blood vessels. When they reach the site to be examined, they strongly reflect the impinging sound waves, making blood vessels clearly evident and unmistakably displaying the tortuous patterns of malignancy. Studies show that this enhancement substantially increases the accuracy of ultrasound in assessing breast and prostate cancer, and in examining the kidneys.

Bubbles also contribute to ultrasound therapy. Like light waves focused through a lens, high-frequency sound waves can be concentrated to deliver energy to a given bodily site. One application is lithotripsy, the use of ultrasound to treat kidney stones. These mineral accretions form in the kidneys when the urine is chemically unbalanced, seriously impairing kidney functions and causing severe pain. The stones can be surgically removed, but focused ultrasound is less traumatic to the body.

When the sound is applied in repetitive bursts at high intensity, it breaks the stones into pieces small enough to be eliminated naturally in the urine. Much of the pulverizing action comes from strongly vibrating bubbles that the sound pulses induce, which also show promise in breaking up blood clots.

Foam plays other versatile roles in biomedicine that depend on its ability to fill a volume and introduce a large surface area, thus effectively combining or distributing active agents. As noted earlier, the chemist Percy Lavon Julian improved the production of synthetic male and female hormones when he processed soybean oil as a foam rather than a solid block of material. And while foam may aid conception in the Japanese quail, it also serves as a contraceptive method for humans.

The history of contraceptive foam illustrates how medical techniques take on political meaning. This foam became part of Margaret Sanger's long campaign to provide women with access to birth control methods. Sanger, born in 1879 as one of eleven children, attributed her mother's early death to the strain of childbearing. That, and her nursing experiences on New York's Lower East Side, convinced her that the means for birth control (a term she invented) should be widely available. Her crusade, including the establishment of the first U.S. birth control clinic in 1916, violated laws and norms of the era. Sanger served jail time, while her views on sexual freedom (she reputedly had liaisons with H. G. Wells and the sexologist Havelock Ellis) stirred considerable controversy.

Nevertheless, by 1938, legal bans on birth control had been eased and it had been endorsed by the American Medical Association. But Sanger had not found a method suitable for poor or poorly educated women—ideally, cheap, reliable, and not requiring extensive instruction or a visit to a doctor. One method Sanger tried was a foam powder that killed sperm. The technique was easier to use than a diaphragm, which must properly cover the cervical entrance to be effective. The powder was impregnated into a small sponge. Attached to a string so it could be removed, the sponge was inserted into the vagina, where the powder foamed up like soap, distributing its spermicidal power and also blocking the entrance to the cervix.

Inadequate testing and other factors prevented foam powder from fulfilling Sanger's high expectations. But spermicidal foam is now widely used as a nonprescription method of birth control that is simpler than the powder-and-sponge approach. The foam, with the consistency of shaving cream, is inserted with an applicator. It is effective immediately and is considered medically safe; still, it has drawbacks. To make sure its bubbles have not collapsed, intercourse must occur within twenty or thirty minutes of application—a loss of spontaneity many find undesirable. And foam is less effective than other methods of birth control. Although in theory it yields a failure rate of just a few percent, realistic figures suggest that a couple using only this method stands a 25 percent chance of impregnation within a year.

HARMFUL VOIDS

From their biomedical uses to their role in the functions of life, foamy structures fulfill important and even essential purposes in our existence. But bubbles and foam can also be biologically harmful. A large air bubble introduced into the blood can block an artery or vein. This obstruction, called an embolism, may occur during surgery or when lung tissue is traumatically ruptured, and can damage the heart or brain to the point of death. The deliberate injection of air into the veins has been implicated in mercy killings, as in a 1950 case when a New Hampshire doctor was prosecuted—although unsuccessfully—for the murder of a dying cancer patient.

Some diseases arise from the insertion of many bubbles filled with gas, or with liquid (in which case they are called vacuoles), into an organism, disrupting its functions. The best-known of these ailments, variously called decompression sickness, caisson disease, or the bends, has been called a disease of bubbles. It can be experienced by fliers who reach high altitudes as well as by divers who spend lengthy periods of time deep underwater in diving suits, by scuba divers (one recently died of the disease after exploring the wreck of the liner *Andrea Doria* off Nantucket Island), or by people who work in the pneumatic caissons used in underwater construction.

A caisson is a large cylinder like a tin can with its bottom removed. It is sunk into a river or harbor, pumped dry, and filled with high-pressure air to keep water out, allowing workers to function on the bottom without breathing equipment. Caissons were placed in the East River to lay the massive piers for the Brooklyn Bridge, the great nineteenth-century construction project, whose chief engineer was himself a victim of caisson disease. Washington Roebling was so crippled after work on a previous underwater project that he could not visit the bridge; instead, he oversaw its construction through binoculars from his home overlooking the site.

For workers in caissons, divers, and aviators, the common element bringing on decompression sickness is a rapid change in pressure. At ordinary atmospheric pressure, the mostly liquid tissues of the body contain the gases of the air in dissolved form. Under high pressure deep underwater, especially over long periods, the tissues absorb additional gas. If a diver rises slowly to the surface, there is time for these gases to diffuse from the tissues into the blood, from which they are expelled by normal breathing. But if the ascent is too rapid, the gas foams into bubbles before it reaches the blood, like a can of soda when its top is popped. Similar bubbles can develop within the body of an aviator in an unpressurized aircraft that rapidly rises above 18,000 feet in altitude, as the atmospheric pressure changes from normal to low (rather than from high to normal).

The bubbles, which contain mostly nitrogen, can have grave consequences. Foaming up within muscles and joints, they cause severe pain and make it impossible to straighten the limbs, the reason the affliction is called the bends. If they accumulate in the respiratory system, breathing becomes difficult. Worst of all, bubbles massing in the brain or spinal cord cause serious convulsions (known as *divers palsy*) or paralysis and sensory distortion (called *divers staggers*). Death may result, but decompression sickness can be treated if the victim is immediately placed in a high-pressure chamber, which forces the gas back into the tissues, and then slowly decompressed to properly eliminate the nitrogen. This relieves the pain and often reverses the ill effects, although nerve damage may be permanent.

The disease of the bends is caused by gas bubbles. In other afflictions, the appearance of bubbles or voids where they should not occur is symptomatic rather than causal. The formation of bubbles marks the most virulent form of gangrene, which is the death of soft tissue when its blood supply is inadequate. This type, gas gangrene, is caused by a particular bacterial infection in deep penetrating wounds. The bacteria exude a highly lethal toxin that forms numerous gas bubbles under the skin.

Another example is the progressive and fatal neurological disorder called Creutzfeldt-Jakob disease (CJD). As Richard Rhodes puts it in his book *Deadly Feasts: Tracking the Secrets of a Terrifying New Plague,* CJD is a disease of holes—that is, holes in the brain. It is one of a cluster of fatal human and animal afflictions whose underlying relations took decades of medical detective work to untangle. They include kuru, another brain disease, found among the Fore people of New Guinea who are said to spread it by cannibalism; scrapie, so-named because the sheep whose brain it affects scrape off their wool in search of relief from the terrible itching it engenders; and mad cow disease, first observed in British dairy herds in 1985, which affected some 160,000 cattle in Britain alone in the following decade. Some researchers report evidence that mad cow disease can be transmitted to humans.

All these diseases are forms of transmissible spongiform encephalopathy, or TSE. *Encephalopathy* means "inflammation of the brain" (it is rooted in the Greek word for brain), and *spongiform* means what it sounds like: The brain becomes filled with so many holes that it becomes spongy, so many that it may lose half its normal volume. One would expect this bizarre appearance to show up in autopsies, where the brain is sectioned into slices and examined under a microscope to search for anomalies. As Rhodes relates, spongiform holes were often seen by pathologists studying CJD. But they were discounted through the "excusable reluctance" to ascribe significance to "nothing," and because they might have been artifacts caused by the pickling methods used to preserve brains. Eventually the universal spongiform nature of TSE became established, however. In all cases, the characteristic voids arise after infected neurons in the brain die and are cleared away by normal bodily activity.

The agent that kills neurons in TSE is extraordinarily hardy, resisting heat, ultraviolet radiation, alcohol, and formaldehyde. According to the World Health Organization, it may be a virus, the extremely small infectious agent responsible for diseases from the common cold to AIDS (caused by HIV, human immunodeficiency virus). But TSE may be caused by a different agent, a self-replicating menace dubbed a prion. That name, for "proteinaceous infectious agent," was bestowed by Stanley Prusiner, awarded the 1997 Nobel prize in medicine for finding this proposed new vehicle for disease. A prion is a protein of unknown function that resides on the surface of brain cells. Prusiner showed that this complex molecule can adopt one particular form that is toxic. Like a single falling domino that mows down a whole chain of other dominoes, this rogue version flips other prions into abnormality and propagates neural destruction through the brain. Considerable evidence has been gathered to implicate prions in TSE, but some scientists believe it is still too early to eliminate viral causes.

Other diseases also reveal their presence through a telltale foamy structure. The oldest of these, leprosy, has been known for over two millennia. No longer the scourge it once was, it still affects up to 12 million people worldwide, mostly in Southeast Asia, Africa, and South America. It is caused by rod-shaped bacteria that are carried through the body by amoebaelike macrophage cells, the very units that are meant to guard the body by engulfing and digesting invading cells. As many as 300 bacteria cram themselves into a single macrophage, swelling it like a balloon, and creating a distinctive foamy appearance visible under a microscope.

Among viruses, one particular family is specifically named for the foamlike appearance its members induce in cells. These foamy viruses (also called spumaviruses) have been studied for over forty years, after they were first observed in cultures taken from rhesus monkeys. In the laboratory, they fill infected cells with bubblelike vacuoles full of virions, that is, functional pieces of the virus, as can be seen under high magnification.

One form of foamy virus, called simian foamy virus or SFV, is widespread among primates including baboons and in other animal species. There is also HFV, human foamy virus, which has been isolated from human cells. But HFV is not known to cause any specific human illness; in fact, it has been called "a virus in search of a disease." Still, foamy viruses attract researchers because they share characteristics with the HIV virus. Both are highly persistent and both use the same mode of reproduction in the cells that they enter. That feature has led some researchers to suggest that if HFV is truly harmless, it could become a useful biomedical tool that could deliver specific genetic information to human cells. But whether the foamy nature of the virus would play a role in this use is unknown.

MATHEMATICALLY PERFECT LIFE

All the biological foams and foamlike systems discussed so far have structures that are not perfectly regular, as is easy to observe. At the microscopic level, Robert Hooke noticed immediately that cells are not identical. The striking images in D'Arcy Thompson's *On Growth and Form* further illustrate that biological cells are not perfect clones of each other, nor are they made with squared-off right-angled corners; as Thompson points out, these are forbidden by Plateau's rules for foam.

Except for the regular hexagonal cells in a honeycomb made by bees, the matrices of life are complex and irregular. That intricacy is responsible for the fascinating variety of living shapes but makes it difficult to examine how interactions among cells lead to the behavior of tissues, organs, and organisms. This theme of relating complex high-level behavior to simpler linked entities appears in other arenas, such as electronic computation, which depends on a network of electronic devices; and the study of human thought processes, which depend on billions of linked neurons in the brain. Among the most fascinating of these pursuits are artificial intelligence (AI), the search to create machines that go beyond ordinary computers to "think" at least as well as humans; and artificial life (ALIFE), ambitiously defined as "the construction of living systems out of nonliving parts."

The trick in these investigations is to find a cellular model that captures the essential elements of life and other complex systems, yet which is simple enough to be manipulated; that is, to create a mathematical foam that can be explored to yield new insights. In the form called a cellular automaton, such an abstract foam was introduced in the 1940s and 1950s by the Hungarian-born mathematician John von Neumann and the Polish-born mathematician Stanislaw Ulam, as a model that can simulate complex processes including some remarkably lifelike ones.

Von Neumann is best known for his theory of games, which has influenced economic thinking and strategic planning. He also contributed to quantum physics, weather prediction, and the theory of computation. Ulam also had a variety of interests and worked with Edward Teller to develop the hydrogen fusion bomb at the Los Alamos Scientific Laboratory. Von Neumann was especially interested in cellular automata that were capable of reproducing themselves. He thought that they could embody a level of complexity approaching that of biological systems and could also lead to ultrareliable self-repairing computers.

The geometric manifestation of a von Neumann-Ulam cellular automaton is simply a regular grid of cells, an idealized foam. The cells could be of any shape but are usually taken as identical squares in two dimensions, like a checkerboard. These right-angled shapes are easy to visualize or display on a computer screen, while supporting the key feature of cellular automata that gives them the power to generate intricate behavior: The condition of any given cell depends on the state of the surrounding cells.

That apparently simple requirement creates unexpectedly rich outcomes. A striking example that is easy to appreciate comes in the computer amusement called the "Game of Life," which perfectly illustrates how cellular automata work. It was first described in *Scientific American* magazine in 1970. Thirty years later, the game has become a classic, retaining a coterie of enthusiasts who follow its ramifications and present them over the Internet, and serving as a model for a variety of serious applications.

The Game of Life plays itself out on a two-dimensional grid

of square cells. Each cell is surrounded by eight others, like any checkerboard square. A given cell is either occupied or not by a living creature, which can be thought of as a unicellular organism, a cyberparamecium or cyberamoeba. These organisms are born, live, and die—in short, evolve—according to the following rules. If an occupied cell has no neighbors or just one neighbor, its organism dies of loneliness. With four through eight neighbors, the organism dies through overcrowding; with the happy medium of two or three neighbors, the organism survives. A new organism is born whenever an unoccupied cell is surrounded by exactly three occupied cells; that is, within the rules of the game, this is the sexual combination that produces offspring.

These simple regulations create astonishingly lifelike behavior. One real-time Game of Life display on the Internet shows a busy, buzzing population occupying a flat universe of over 65,000 cells, that is, 256 cells each in the vertical and horizontal directions. Starting with just a few organisms, and following them through generation after generation as the rules are applied, one sees multicell beings grow and die, coalesce and break up, and move rapidly across their abstract universe, sometimes colliding and combining into new creatures. It is the cybernetic equivalent of the lively moving "animalcules" Antonie van Leeuwenhoek saw through his microscope in the late seventeenth century.

The Game of Life handily illustrates how an abstract cellular geometry can create or simulate complex behavior. The patterns generated by cellular automata have been connected to actual biological systems such as leaves and branches, skeletal structures, and designs on seashells. In addition, the principles of cellular automata have proven useful in solving certain mathematical problems and have been applied to instrumentation that automatically counts and analyzes blood cells at high speed, and to the design of complex computer circuits. Still, we will not be seeing the mathematical foams of cellular automata turning into ALIFE any time soon.

It would take a book the size of Thompson's *On Growth and Form* to do justice to all the foams and foamy structures with biological meaning. We have not examined, for instance, the cellular arrangement in coral; or the role of alveoli in lactation; or the radiolaria, tiny marine organisms whose foamlike clusters of cells resemble fantastic three-dimensional snow crystals. Even so, it is clear that foam and foamlike structures are significant for life, from the microscopic cellular level to the reproductive behavior of fish to the physical arrangement of the human body.

Other foams are significant in the natural nonliving parts of our planet. In the next chapter, we examine pumice, found in the rocky bulk of the Earth, and sea foam, which is spread widely over the oceans.

six

EARTHLY FOAM

VOLCANOES, OCEANS, AND CLIMATE

The Earth displays natural foams which arise from the physical processes that have sculpted our planet, and still do. Sea foam, dynamically born as air and water beat together to create immense numbers of bubbles, represents the impressive power of breaking waves and stormy seas—it is so prevalent that it can even be detected by Earth-scanning satellites in space. By comparison, the foamy rock called pumice (the name is related to the Latin root that underlies the Italian *spuma*, meaning "foam") seems dull and static. Each of its bubbles remains fixed within a solid matrix for the lifetime of the foam, which can stretch over millions of years. Yet this stable structure originates in a process as violent as any on Earth: the explosive eruption of a volcano, which releases gases that turn molten rock into an airy solid.

Apart from their common foamy geometry, pumice and sea foam could not be more different, but there is one great similarity. Both are connected to the nearly 5-billion-year history of the Earth: pumice because the volcanoes that spew it out have long shaped the Earth and its atmosphere; sea foam because it represents and controls immense exchanges of matter and energy among sea, air, and land that affect our planet and its life. In fact, as we learn more about the mighty systems that comprise the Earth—its rocky crust, the lithosphere; its sheath of gases, the atmosphere; and its seas and oceans, the hydrosphere—the more we appreciate that they must be treated as an interrelated whole.

The relations extend even farther, for life itself, the biosphere, depends on the physical characteristics of our planet, which we now understand have been influenced in turn by the presence of life. Few scientists would accept the extreme form of this view, the Gaia hypothesis, put forth by the British physicist James Lovelock and the American microbiologist Lynn Margulis. They propose that the biosphere, atmosphere, hydrosphere, and lithosphere form an integrated global organism that controls its own processes. But increasingly we recognize relationships both obvious and subtle among earth, air, water, and life. This complex story, which stretches from the origins of the Earth to contemporary issues such as climatic change and depletion of the ozone layer, includes important chapters about natural foams made in the sea and emitted from volcanoes.

VULCAN'S FOAM

The volcanoes that make the solid foam pumice have inspired awe for millennia. The ancient Greeks wrote about eruptions from Mount Etna in Sicily, which they took to be the workshop of Hephaestus, the god of fire. The Romans transmuted Hephaestus into their own fire god Vulcan, also associated with volcanic destructiveness. Nowadays scientists have replaced these myths with solid knowledge of volcanic origins and behavior, although they do not yet understand everything about volcanoes.

In a volcanic eruption molten rock pours through a vent in the Earth's crust more or less explosively, forming lava and other products that may include pumice. The rock emerges at temperatures up to 1,200 degrees Celsius (2,200 degrees Fahrenheit). It is called *magma* (the word comes from the Greek root meaning "to knead," a reference to its plasticity) and originates in the Earth's mantle—the semifluid interior shell, 1,800 miles thick, surrounding the Earth's central iron core and surrounded in turn by the rocky outer crust. The crust is on average only 12 miles thick, a minute fraction of the 8,000-mile diameter of the Earth. When hot magma breaks through that rocky eggshell, a volcano may erupt.

Volcanoes occur largely at the boundaries between tectonic plates—the dozen or so enormous rafts of rock, and several smaller ones, that constitute the crust and upper mantle, support the oceans, and form the continents. These plates float atop the semiliquid mantle and collide, drift apart, or slide past each other like a slow and colossal game of bumper cars played over millions of years. Many volcanoes and earthquakes occur where one plate rides up over another (which happens at the imperceptible rate of inches per year). One massive volcanic belt, the Ring of Fire, surrounds the Pacific Ocean along the edges of its underlying Pacific Plate. It includes Mount Saint Helens in Washington State, which erupted violently in 1980. Mounts Etna and Vesuvius are part of a second group at a juncture of tectonic plates under the Mediterranean Sea. (Some volcanoes are also found far from plate boundaries, as in the Hawaiian Islands.

They are thought to form above "hot spots," where plumes of magma rise through the mantle as warm air rises.)

The ferocity of any particular volcanic eruption depends on the properties of the escaping magma, which contains gases formed from the elements within the Earth. Like carbon dioxide in a can of soda, the gas is dissolved rather than in the form of bubbles; it is constrained by pressure, in this case the immense weight of the overlaying rock. But just as a soft drink fizzes when it is uncapped, the magma froths into bubbles as it rises to regions of lower pressure. If the magma has a fluid, free-flowing composition, the bubbles escape, often producing an initial explosion; then the remaining magma emerges and cools into lava. But if the magma is thick or viscous, the gas cannot readily break free. Pressure builds up until a major explosion results, sending volcanic products into the air. Among these is pumice, which is magma that has reached the surface, then cooled and set so rapidly that its bubbles became trapped before they could dissipate. This process can produce so many gas-filled voids that the result is a solid foam light enough to float on water.

Bill Size, a geoscientist at Emory University who has examined pumice around the world, made it possible for me to test that surprising ability to float. A dark gray piece of pumice he provided had obvious voids but retained the heft of ordinary rock and sank just as quickly. A dirty gray-white sample with bigger cells was noticeably lighter but did not quite float. A tan piece, however, felt no heavier than a bit of foamed plastic and bobbed along merrily, only half submerged, in a bowl of water.

This floating rock was mentioned in the influential and encyclopedic *Natural History*, written by the Roman scholar Pliny the Elder (to distinguish him from his nephew Pliny the Younger). Born in 23 C.E., he served as a cavalry officer in Africa and Germany, then returned to Rome to take up study and writing. *Natural History* is a compendium of what was then known about the natural world, including cosmology, zoology, botany, medicine, and mineralogy. In that last section, Pliny refers to "a stone from the island of Syros [that] floats on the waves," and later describes this pumice stone. "The test of its quality," he writes, "is that it should be white, very light in weight, extremely porous and dry, and easy to grind." He goes on to list its uses: As an abrasive, it smoothes the edges of book rolls and acts as a depilatory for men and women. In powder form it is a remedy for painful teeth, and when ingested, increases the capacity of drinkers to hold their wine.

Nowhere, however, does Pliny give the origins of this miraculous stone. That carries a certain irony, for later Pliny witnessed the source of pumice, a violent volcanic eruption—an encounter that ended in his death. By a quirk of fate, his nephew Pliny the Younger wrote an account of that very event. It was the world's first detailed portrait of a major volcanic explosion, the eruption of Mount Vesuvius in 79 C.E. that buried the Roman city of Pompeii and killed its people along with the elder Pliny. The particular type of eruption Vesuvius displayed is now called *Plinian* in his honor. It is the most savage type known to volcanologists; gases blast straight up out of the magma like the launch of an enormous space shuttle, carrying fragments that fall as ash and pumice.

At the time of the eruption, the elder Pliny commanded the Roman fleet based in the Bay of Naples. According to his nephew, as his uncle neared shore aboard ship, "cinders . . . grew thicker and hotter the nearer he approached, fell into the ships, then pumice stones too, with stones blackened, scorched and cracked by fire."

Later, when Pliny and his companions reached land, "in the open air they dreaded the falling pumice stones, light and porous though they were. They tied pillows upon their heads with napkins; and this was their whole defense against the showers that fell around them."

Pliny the Elder did not die of falling pumice, but of asphyxiation, as he breathed noxious fumes from the eruption. So heavy was the fall, however, that Pompeii and its surroundings were buried in up to twenty-three feet of *pumice lapilli*—small stones made of pumice—and volcanic ash. In all, Vesuvius discharged about a cubic mile of pumice.

Not every volcanic event makes pumice, but two of the fiercest eruptions in modern times ejected huge quantities of the material. The first was the cataclysmic 1883 explosion of Krakatau in Indonesia. It killed upwards of 35,000 people and threw so much gas and debris into the air that sunsets and the climate were affected for a year afterward. The surrounding seas were covered in five feet of pumice, thickly enough to obstruct ships, and floating rafts of the stuff traveled across the Indian Ocean to Africa. Over a century later, in 1991, Mount Pinatubo in the Philippines sustained its own Plinian eruption. Although Pinatubo had apparently been dormant for over 600 years, its eruption was the most powerful since Krakatau, ejecting a twelve-mile-high column of material that produced global climatic effects, and generating up to three cubic miles of ash and pumice.

These cubic miles of puffed-up rock are impressive signs of volcanic power, and the void-filled geometry of pumice helps us analyze the behavior of volcanoes. That tells us about the interior of our planet. Surprisingly, it also extends our knowledge of how the Earth's atmosphere evolved to its present unique (as far as we know) oxygen-rich life-supporting composition—a process that depends partly on the gases ancient volcanoes emitted. Fuller knowledge of volcanism also allows us to cope better with its adverse effects. The sulfur dioxide gas that volcanoes emit has harmful worldwide effects on climate and atmosphere. In addition, there are the immediate lethal effects of eruptions. They are estimated to have killed 200,000 people over the last 500 years, and still take their toll; 57 people died in the eruption of Mount Saint Helens. The ability to predict eruptions has improved through the modern study of volcanic events, but forecasting these deadly occurrences is not yet an exact science.

Pumice actually records volcanic activity through its scouring action, which leaves marks that can date an eruption if it is relatively recent. An eruption showers vegetation with falling pumice, which strips leaves or needles from trees through the abrasive action of its jagged edges. That may not kill trees outright, but pumice-blasted trees hardly grow for several years. The result is a series of extremely narrow, highly noticeable tree growth rings, which can be used to count back from the present age of the tree to the year of the eruption.

The voids in pumice may also hold enduring records of older eruptions, which can be read through the technique of radioactive age dating. Many rocks and minerals contain radioactive atomic elements, such as uranium, whose nuclei spontaneously emit some of the protons and neutrons that comprise them. That process changes the nucleus into some other stable non-radioactive element—uranium eventually turns into ordinary lead—at a rate that is precisely known. If the stable atoms accumulate in the rock (which depends on the element, the type of rock, and its environment), then careful measurement of their number relative to the remaining number of radioactive atoms gives the age of the rock. The radioactive decay of uranium into lead has been used to establish the age of the Earth as 4.6 billion years.

Other radioactive sequences work well for less awe-inspiring time spans. One in particular, the decay of a radioactive isotope of the element potassium into the inert gas argon, can be tracked by examining crystals preserved within pumice. In this way, pumice has been used to date a major volcanic event that took place in historic time: the very eruption of Vesuvius that Pliny the Younger reported as occurring in 79 C.E. The team of

Italian and American scientists that carried out this study did not question Pliny's date; rather, the plan was to use it to test the potassium-argon method. The researchers examined white pumice recently excavated from Oplontis, the villa built at Pompeii for Poppaea, wife of the emperor Nero. The villa was buried in pumice that held crystals of the mineral sanidine, which is rich in potassium. In 1997, after elaborate measurements of the argon resulting from the decay of the potassium, the team concluded that the pumice had formed 1,925 years earlier, agreeing within experimental error with Pliny's date of 79 C.E., that is, 1,918 years earlier.

PRESERVING BRIMSTONE IN PUMICE

The foamy geometry of pumice also safeguards other volcanic products that affect us today and reflect the history of the Earth and its life. Among these are chemical compounds of sulfur, which has long been associated with the infernal regions under its old name, brimstone. Sulfur is essential for life. It is found in every living organism, which it enters as part of a cycle that interchanges sulfur among atmosphere, rocks, oceans, and living things. Present-day volcanoes are a global source of sulfur, injecting it into the atmosphere as gaseous sulfur dioxide, SO_2 (along with carbon dioxide, water vapor, and other gases). We do not know, however, whether ancient eruptions also ejected sulfur, nor how its presence or absence affected the Earth's early atmosphere and the origins of life.

Today, SO_2 has serious adverse effects when it combines with water vapor in the air to make sulfuric acid, one component of acid rain (the other is nitric acid). The outpouring of SO_2 from an eruption can cause acid rain hundreds of miles away. A bigger issue is that a powerful Plinian explosion blasts tremendous amounts of SO_2 into the stratosphere. That turns into a fine mist of sulfuric acid droplets that spreads around the world, propelled by the high winds at that altitude. One extreme result may be seen on the planet Venus, the brightest object in the sky (after the Sun and the Moon) because it is blanketed with clouds that brilliantly reflect sunlight. The clouds are made of sulfuric acid, which almost surely comes from volcanic activity on Venus.

On Earth, the 1991 explosion of Mount Pinatubo expelled 20 million tons of SO_2. A month later, sulfuric acid was covering much of the Earth, blocking part of the Sun's radiant energy. Within two years, this cloud had cooled large parts of our planet by up to 0.5 degrees Celsius (0.9 degrees Fahrenheit), a change big enough to correlate with notable climatic events like harsh winters.

In another worrisome effect, the acid cloud depletes atmospheric ozone, the oxygen molecules that absorb harmful ultraviolet light from the Sun. A thinner layer means that more ultraviolet light reaches the Earth's surface, which leads to higher rates of skin cancer—now the most prevalent form of cancer—and cataracts of the eyes.

Tracing how sulfur gets from the Earth's interior into the atmosphere is essential to understand climatic behavior. One important aspect is the relation between the emitted SO_2 and the magma it came from. That relation could be obtained by examining the solid products of an eruption, but unfortunately, sulfur compounds dissolve in water. Especially for old eruptions, where the residue has seen much weathering and precipitation, this can mean a lack of readable clues.

Pumice, however, preserves the clues, a point made by researchers who examined the pumice from Mount Pinatubo. It contained crystals of anhydrite, a sulfur-bearing mineral that ordinarily would have dissolved, but which remained in useful condition within the pumice. Volcanologists plan to further pursue the clues by examining pumice from ancient eruptions to learn how much sulfur was emitted then. Such results would be a step toward establishing the role of volcanoes in the long-term development of our atmosphere, and perhaps of life.

OTHER VOID-FILLED ROCKS

Pumice is not the only foamlike rocky structure. Smaller concentrations of voids (or vesicles as geologists call them) are often found in basalt, a dense and widespread gray or black volcanic rock. Another vesicular volcanic rock called phonolite occurs in the 865-foot-tall Devils Tower National Monument in Wyoming, made notorious in the 1977 movie *Close Encounters of the Third Kind*. This tree-stump-shaped mass contains vesicles whose density varies from its top to its bottom, indicating changing conditions as it was formed. In general, the nature of the voids in volcanic rock indicates where it was made. Basalt found on the deep ocean floor, formed under high pressure, is only about 1 percent porous, with small bubbles typically ten micrometers across. When vesicles of similar size and concentration are seen in basalt found on land, that may indicate the rock originated beneath the sea.

Vesicular rocks also exist elsewhere in the solar system. Astronauts David Scott and James Irwin picked some up from the surface of the Moon during the Apollo 15 mission of 1971. These rocks did not come from volcanoes, but from another violent process, the formation of craters by meteorites. On Earth, most would-be meteorites burn up in our dense atmosphere. On the airless moon, incoming chunks of rock hit the surface at speeds of thousands of miles per hour. They carry enough energy to turn lunar rock molten, splashing it up to form the circular wall that defines a crater, and throwing out molten chunks called *ejecta*, which trap bubbles of gas when they solidify. The rocks Scott and Irwin found are thought to represent ejecta from craters far distant from the astronauts' landing site. Rocks with a vesicular appearance also exist on Mars, as shown in photographs of the surface taken during NASA's Pathfinder mission in 1997; these rocks could also represent ejecta from impact craters, or they might have been formed by volcanic action.

OCEANS OF FOAM

Although rocky foams like pumice and ejecta are not unique to Earth, sea foam seems to be. As far as we know, our planet is the only place in the universe with open water that sustains foamy waves or white water. For sheer global impact, pumice does not begin to approach the influence of sea foam, which is widespread over the oceans.

Seawater contains surfactants, and so sea foam lasts for a while. We should expect no foam at all in pure water, which lacks surfactants, but foam appears wherever water, even fresh water, mixes vigorously and continuously with air. My backyard is graced with a man-made stream where water runs down an inclined bed lined with rocks, a small flow that develops inch-high flurries of white effervescence. Truly imposing examples appear in the world's great cascades such as Africa's Victoria Falls between Zambia and Zimbabwe. There the descending water fills the deep pool of the Boiling Pot below the falls with churning foam. No surfactant is needed because the dynamic mixing of air into water makes new bubbles fast enough to replace old ones as they die. Even in the ocean, air and water must first beat together to make foam. But no matter where or how the foam is made, it is white for the same reason soap foam is white: It is filled with bubbles that scatter light.

The Boiling Pot is a spectacular example of foam made from agitated air and fresh water, but on the global scale, sea foam is

A breaking wave churning up foam.

more important than all the freshwater foam in the world. It arises in different ways. Far out to sea, the foam is a by-product of how wind makes waves. Closer to the shore, it is formed as waves encounter the sea bottom, which makes surf. The roaring energy of surf is obvious, and it is easy to see how it changes the edges of continents by eroding beaches; what is less obvious is that whitecaps have, by far, the greater global impact.

Whitecaps ride atop waves, and waves are made as wind piles up water ahead of itself. The interaction is complex, depending on the strength of the wind, how long it has been blowing, and the amount of sea over which it has already passed, called the *fetch*. Generally speaking, however, if the wind blows with enough force (the critical speed is usually quoted as around fifteen miles per hour, or in nautical terms, 13 knots, although at least one study puts it lower, at seven to eight miles per hour), it pushes the water at the top of the wave so hard that it moves faster than the body of the wave. That makes the crest topple over; as it does, it mixes with air to form bubbles that make a line of white foam, a whitecap.

The number of whitecaps in a given area of ocean, or equivalently the fraction of sea surface covered by their foam, correlates with the speed of the wind over the sea. Long before oceanographers quantified this relation, sailors knew that the sea is foamiest in the southern latitudes called the roaring forties, that is, the 700-mile-wide ring around the Earth that covers the region forty to fifty degrees south of the equator. There the winds blow steadily from the west; there the ocean, nearly uninterrupted by land around the whole girth of the world, builds up enormous foamy waves driven by those winds blowing freely over great reaches of sea.

We now observe the roaring forties in perfect safety from vantage points in space. A spectacular bird's-eye view comes from the Earth-orbiting TOPEX/Poseidon satellite, launched in 1992. While it tracks over the globe, the satellite shines radar beams at the surface of the sea. As these beams interact with the water and reflect from it, they change according to the height and direction of motion of the waves. The satellite senses the reflected beams and transmits the data they carry back to Earth. As a final step, the information is turned into images showing wind speed and wave height for all the world's oceans, which are displayed on the Internet as striking color-coded global maps. Cool blues and greens represent gentle winds and calm seas, whereas yellow, orange, and red indicate progressively higher winds and bigger waves. Within this scheme, the roaring forties are consistently a blazing yellow-orange with spots of red, dramatically showing that wind and wave are greatest there. In the fall of 1998, for example, wind speed reached thirty-five miles per hour and waves were as high as twenty feet.

It wasn't always like this. A hundred years ago, Captain Joshua Slocum saw the power of those foamy seas from a most vulnerable place, the deck of his small wooden sailboat *Spray*. In 1895, Slocum set out from Boston to sail around the world alone, becoming the first person to do so when he returned in 1898. An immensely experienced seaman who had been sailing since the age of sixteen, he carefully planned his circumnavigation to evade the dangerous roaring forties. But he could not avoid dipping into them as he rounded the Cape of Good Hope and Cape Horn, at the southernmost tip of Africa and South America, respectively.

At Good Hope, wrote Slocum, the *Spray* "was trying to stand on her head" and ducked him under water three times in as many minutes. But in his estimation, the seas off Cape Horn were the roughest in the world. In the darkness of approaching night and a gathering storm, all he could see was the "gleaming crests" and "white teeth" of the waves. There he encountered what Charles Darwin, voyaging in the same area on HMS *Beagle* while pondering the theory of evolution, called the Milky Way—a region where foamy wind-driven waves became yet frothier and whiter as they dashed brutally against sunken rocks. Wrote Darwin, "Any landsman seeing the Milky Way would have nightmare for a week." So would any seaman, added Slocum, calling the experience "the greatest sea adventure of my life."

For all the drama, mariners make pragmatic use of sea foam when they employ the Beaufort scale, a method of estimating wind speed by observing waves and foam that was devised by the British naval officer Francis Beaufort. He was an uncommonly able draftsman and chart maker who became hydrographer of the Royal Navy. Before the invention of devices to measure how fast the wind blows, the scale Beaufort developed in 1806 was a step toward the accurate determination of wind speed for sailing men-of-war.

Beaufort's original scale was graduated from winds of force 0, "calm," to those of force 13, "storm." Later the gradations were expanded and matched to descriptions of the state of the sea. A force 6 wind is a "strong breeze," blowing at twenty-four to thirty miles per hour (22 to 27 knots), and producing "white foam crests." A force 8 wind, a "gale," moves at thirty-eight to forty-six miles per hour and makes foam that is "blown in well-marked streaks along the direction of the wind." Winds of force 12 to 17 blow at seventy-three miles per hour or more and are classified as "hurricane," where "the air is filled with foam and spray; sea [is] completely white with driving spray."

Scientists have further refined the connection between wind speed and foaminess, although determining it with mathematical precision has not been easy. It is difficult to make accurate measurements at sea and to simulate whitecaps in the laboratory. Still, there is scientific consensus on how strongly the degree of foaminess depends on the wind. The mathematical relation shows, for instance, that an increase in wind speed from fifteen to twenty miles per hour nearly triples the percentage of sea surface covered by foam.

Using the average sea-level wind speed over all the Earth's oceans, the relation predicts a typical whitecap coverage of 2 to 3 percent of the ocean's surface. Since the oceans cover 70 percent of the Earth, that means at least 3 million square miles of sea surface, the area of the United States, is fizzing with bubbles at any given time. The mathematical relationship also confirms that the amount of foam coverage varies sharply over the globe, being greatest in the roaring forties. And it predicts that 100 percent foam coverage, a totally white sea, occurs at wind speeds of eighty miles per hour, consistent with the Beaufort description of "hurricane" conditions at comparable speeds.

It is easy to see why sailors think about sea foam and the dangers it represents, but why should the rest of us, mostly landlubbers, be concerned? The surprising answer is that whitecaps have significant impact on the Earth, a fact which is beginning to be widely appreciated. Sea foam affects how the Earth interacts with radiant energy from the Sun, part of which is absorbed by the Earth's air, land, and sea, and part is reflected back into the atmosphere or into space. The difference between the energy absorbed and the energy reflected has a large effect on Earth's temperature and climate. Sea foam, which is white, contributes to that balance because it reflects a larger fraction of impinging sunlight than does green or blue seawater. (This also complicates the analysis of data from Earth-orbiting satellites, which use the color of seawater as a guide to the depth of the ocean and an indicator of its plant and animal life.)

The bubbles in sea foam paradoxically gain particular importance just as they die by breaking at the surface, because that is when they spread their contents into the atmosphere and around the world. Sea bubbles come mainly from whitecaps, although rain and other mechanisms also contribute. As discovered in the 1960s and 1970s by Herman Medwin and his students at the Naval Postgraduate School in Monterey, California, the bubbles are too small to see. Typically they measure 100 micrometers in diameter. But these microbubbles add up because they are made in huge amounts, thousands per cubic foot of seawater; and because under some conditions they form in groups called bubble clouds that act together in a coherent way.

Microbubbles are prevalent in what oceanographers call the surface microlayer of the ocean, its top millimeter which contains much of its microscopic life and has been called a last frontier in the study of the sea. It is an air-water interface that is largely affected by its bubbles, which influence how seawater evaporates into the atmosphere and how the gases in the atmosphere dissolve in the sea. And as the bubbles break at the surface of the ocean, they emit tiny plumes of water. These spread the water and what it carries—salts and other inorganic materials, organic materials and parts of marine organisms, even electrical charge—into the air.

Each microbubble launches only a minute amount of material. The total depends on the number of bubbles, which varies with whitecap coverage and therefore wind speed, so the greatest contributions worldwide come from the roaring forties. But even at a comparatively low wind speed of twelve miles per hour (a "gentle breeze" on the Beaufort scale), a square yard of sea surface emits up to nine grams of water into the air per day. This small contribution turns into a stunning amount when multiplied by the surface area of the oceans—billions of tons removed daily, up to a cubic mile of water that enters the unending planetary cycle of rain; replenishment of rivers, lakes, and seas; and rain again. Similarly, although each breaking drop releases a tiny electrical charge like that contained in a few electrons—the smallest units of electricity (except in this case the charge is positive)—in aggregate, that contributes substantially to the electricity in the atmosphere.

Breaking bubbles also distribute what the water carries, removing as much as 1,000 million tons of salt—2 trillion pounds—from the seas yearly, according to one estimate. The salt does not fall right back into the ocean but is left in the atmosphere after the water evaporates, in a finely dispersed form easily spread over the Earth by winds. The salt particles and other materials that the bubbles diffuse into the atmosphere are essential in the world's weather cycles; they form nuclei around which atmospheric water vapor accretes, eventually condensing into rain and the droplets that make clouds.

Salt carried into the atmosphere has other effects, such as the "salt air" that tarnishes automobile chrome near the seashore. Even far inland, however, even in arid deserts, wind-borne compounds from the sea promote chemical reactions that weather rocks. More important yet may be the organic material the bubbles carry in relatively large amounts. Current thinking has it that life began in the seas—perhaps in warm shallows, perhaps in hot thermal plumes deep underwater—which provided an environment for the formation of DNA, chlorophyll, and the other complex molecules of life. If so, it is natural to imagine (and for now, it is only speculation) that the injection of preliving or living matter into the atmosphere helped distribute the conditions for life, or life itself, over the Earth.

SOUNDS IN THE SEA

Sea bubbles affect another feature of the oceans themselves: their natural sounds. For humans, sound is secondary to light; in response to the sunlight bathing the Earth, vision has evolved into our dominant sense. But the undersea world is illuminated by sound, not light. Seawater greedily soaks up sunlight (and every other form of electromagnetic radiation), as you can see when you enter the dim realm a few feet beneath the ocean's surface. Continue down another few hundred feet, and even that faint light is gone. In the deepest ocean trenches, there is only the glowing bioluminescence of living creatures.

These beings live in a dark world, but not a quiet one. Sea-water is highly transparent to sound, transmitting it without losing much of its energy. In one recent test, researchers generated undersea sounds that were received 11,000 miles away, nearly halfway around the world. Sound is a significant mode of sensing and communication for the sea's inhabitants, as in the song of the humpback whales, and the sound waves dolphins use to navigate in the process called echolocation.

We exploit the efficient underwater transmission of sound with sonar, SOund NAvigation and Ranging. Passive sonars use underwater detectors—hydrophones—to listen for ships and submarines. Active sonars imitate dolphins by radiating into the water pulses of sound (the dramatic "pinging" heard in every submarine movie ever made) which are reflected back to a hydrophone when they encounter an object. During the Cold War, the United States constructed SOSUS (SOund SUrveillance System), a vast network of hydrophones deep under the Atlantic and Pacific Oceans, to detect enemy missile-bearing nuclear submarines. Now SOSUS is turned to uses such as tracking whales, and sonar finds schools of fish and explores the sea bottom.

The faint sounds of distant whales, however, or the weak echo from a hostile submarine, can be difficult to discern against the background sounds of the sea—just as it is hard to see the dim flame of a candle against sunlight streaming into a room. Analogously, the background sound permeating the oceans has been called *acoustic daylight*. And that is where bubbles come in: According to researchers like Lawrence Crum of the University of Washington, who has studied bubbles for thirty years, most of that "daylight" comes from the microbubbles formed from whitecaps. These microbubbles are noisy because they oscillate rapidly in size when they are created. The oscillation creates rapid pressure fluctuations in the air trapped in the bubbles, generating sound over a broad range of frequencies.

Noise from bubbles sets an ultimate limit on what we can hear beneath the sea, but it carries information as well. The characteristic sounds of bubbles have been used to track waves and sea currents, and the bubbles that appear when raindrops strike the sea are also useful. The size of the drop sets the size of the bubble and hence the frequency of its sound, so that with educated listening, it is possible to determine how hard it is raining at a distant location in the ocean. Since rain-monitoring stations are sparse over the seas, researchers have used these sounds to help them forecast weather and study global warming. Recent research by Crum shows that even a snowflake falling on the sea—surely among the most gentle impacts in nature—produces a bubble with a characteristic acoustic signature. (Crum speculates that to the acute hearing of a dolphin, a snowfall on the surface of the water might sound like a "huge damn thunderstorm.")

Bubbles may also modulate sounds made by other sources, such as underwater seismic disturbances. This possibility was reported in 1996 by scientists monitoring seismic activity in French Polynesia. Their listening stations recorded mysterious low-frequency underwater sounds at three to twelve hertz (cycles per second), far below what humans can hear. The unusual feature, confirmed by analysis using data from SOSUS, was that these were pure tones unlike any other undersea sounds the scientists had encountered; that is, they contained only a single frequency rather than the mixture of fundamental frequency and overtones that lends character to musical instruments and human voices.

The researchers traced the sounds to an underwater volcano 400 feet below the surface and came up with a possible explanation for their purity. The volcano exudes hot lava that turns water into steam. At great depth, water pressure would prevent the steam from making bubbles; but at a relatively shallow 400 feet, a cloud of bubbles forms between the volcano and the sea surface. The sound waves made by the eruption are modified as they enter the cloud and bounce back and forth between its top and bottom. That generates a resonant sound keyed to the size of the cloud, in the same way a bluegrass musician produces a vibrant sound by blowing into a whiskey jug. The sound from the jug contains a rich mixture of overtones; but the bubbles in the cloud absorb overtones, leaving only a single pure tone. While the researchers caution that this model is unproved, it is scientifically reasonable; besides, it is appealing to imagine a volcano delicately playing its own bubble-filled jug to make a pure tone.

THE SURF ZONE

One need not explore an underwater volcano or brave the roaring forties to see violent foam or rafts of bubbles covering large stretches of ocean. These phenomena also appear, and most dramatically, as waves approach the shore to form surf, the turbulent white-water zone between beach and sea. The white line of foam curving over a surfer riding a wave represents an army of bubbles. Harder to imagine is the fact that those bubbles also make the sound of surf, just as they make the far quieter background noise in the open sea. But bubbles are exactly what makes the roar of surf, according to experts such as Crum. The deep relentless boom of surf comes from galaxies of tiny bubbles madly vibrating in the white water made by the breaking waves. Some vibrate singly, which produces higher pitches, and some vibrate in coherent clouds, making lower tones.

The bubbles in a breaking wave or breaker arise from a difference in speed between the top and bottom of the wave. As the wave progresses along the ocean, what actually flows is energy, not water, which hardly moves in the direction the wave travels; instead, the undulating wave makes water molecules vibrate up and down as it proceeds. If that seems hard to believe, watch a floating buoy when a wave sweeps past. As the buoy bobs up to the crest of the wave and descends into its trough, its position on the surface of the sea hardly changes, just like any individual water molecule.

Nothing limits that up-and-down bobbing in deep water, but as a wave nears shore in shallow water, the sea bottom restricts the vertical motion. That takes energy from the wave and slows its lower portion, while the crest continues at its original speed. In effect, the top of the wave somersaults over the bottom. Depending on the wave and the slope of the sea bottom, the resulting breaker can take any of several forms. Most dramatic is a *plunging breaker*, where the entire leading face of the wave steepens until it is vertical. Then the whole crest curls forward over that vertical wall, crashing down to beat air and water into a froth. A *collapsing wave* is a variant that also curls over and falls forward to make foam, but beginning below the crest. In a *spilling breaker*, foam appears on the crest, then covers the whole leading surface of the wave, followed by water spilling forward from the crest. Only a *surging breaker*, a moving bulge of water that glides up a beach without ever quite breaking, is nearly foamless and noiseless.

The energy evident in foaming breakers is exciting to watch and gives surfers the thrills they love. It has serious effects as well, for the crashing water and the currents it creates shift quantities of sand beneath the water and on the beach. That has long-term geological significance in changing the coastlines of continents, and immediate importance as it erodes beaches. Each year the U.S. Army Corps of Engineers spends tens of millions of dollars replacing beach sand lost through erosion. Often that is a thankless and Sisyphean task, because the new sand quickly washes away, but we know so little about natural beach processes that no one is quite sure where it goes. It may move slightly offshore, forming a wave barrier that helps reduce

future erosion; or it may wash out to deep water, never to be recovered. Without understanding how waves and surf move
sand, we are limited in our ability to control erosion and predict
storm damage.

SANDYDUCK '97

To increase our understanding of the surf zone, scientists conducted the biggest field study ever made of coastal processes,
Sandyduck '97. It was carried out in a natural laboratory, the
Outer Banks of North Carolina. The project's odd name relates
to its base of operations, Duck, North Carolina, where the
Corps of Engineers maintains a research facility that is heaven
for surf watchers—a half mile of beachfront, with a pier, a 140-
foot observation tower, and specialized equipment such as the
strange-looking CRAB, Coastal Research Amphibious Buggy.
Imagine a photographer's tripod stretched to a spidery height
of thirty-five feet, and embellished with a wheel on each leg, an
engine, and a small control cab on top; that's the CRAB, which
splashes far out into the surf zone to put sensors in place and
sample the sea bottom.

Sandyduck took form in June 1997, when scientists installed
acoustic and optical instruments, cameras, and other devices to
observe sand, waves, and surf. Data were gathered the following fall, as over 100 researchers and students monitored thirty
individual experiments. The results illustrate the dual power of
foam and bubbles as agents of coastal change, and as the means
to study that change. For instance, data gathered at Sandyduck
showed that the plunging type of breaker emits a specific underwater acoustic signature characterized by a sharp onset of
sound. This may be a first step toward the remote monitoring
of breaking waves for fundamental understanding, and for applications such as amphibious landings in warfare.

Some scientists analyzed the bubbles themselves. Tim Stanton, for example, an oceanographer at the Naval Postgraduate School who specializes in ocean turbulence, tested a conjecture he developed with Tom Lippman of the Scripps Institute of Oceanography, who videotapes surf to observe its foam. The conjecture was that the bubbles formed by a breaking wave would reflect the degree of turbulence the breaker imparts to the water. Intense turbulence would produce energetic eddies extending deep into the water, which would in turn disturb the bottom sand.

To test this idea, Stanton compared the number of bubbles generated by breakers to the degree of turbulence and the amount of stirred-up sand or sediment. He counted bubbles, or rather the percentage of a given volume of water occupied by bubbles, by measuring how well the seawater carried electricity. Pure water does not carry electrical current, but seawater does, because of its dissolved impurities (this is the reason seawater absorbs electromagnetic radiation including visible light). Mix in bubbles, however, and the flow of electricity decreases, because air does not carry electrical current. By measuring the underwater flow of current, Stanton was able to find the relative fraction of bubbles.

Stanton also used acoustic instruments, including the imposingly named Bistatic Coherent Doppler Velocimeter, which he designed, to determine the turbulence of the water and the concentration of sediment. The electrical and acoustic measurements were made starting at the seabed and working upward toward the water surface in small steps, to measure behavior at different depths. The instruments were mounted on a sled that was towed into deep water 600 feet from shore by the CRAB, and then was slowly pulled back in, gathering results at different locations.

The experiment produced reams of data that are still under analysis, but has already clarified how bubbles transfer energy from breaking waves to water and sand within the surf zone.

The outcomes include spatial profiles of the bubbles made by breakers; that is, how their numbers vary with distance beneath the surface of the water, and with distance from the breaker that made them. And the data seem to show correlations between the number of bubbles and the degree of underwater turbulence. That would confirm Stanton's hypothesis and provide a first step toward the day when it will be possible to simply look at surface foam in the surf zone and relate its character to turbulent behavior deep underwater.

Crashing white surf represents violent sea foam, in vast contrast to the quiet, long-lived foam trapped in pumice. Ironically, pumice begins as a foam of bubbles generated in volcanic violence, yet if those researchers in French Polynesia are right, a volcano under the sea makes fragile bubbles in water as well as long-lasting ones in rock. These and other connections among sea foam and volcanoes, surf and pumice, symbolize the global role of foam and its ties to the innermost and outermost processes of our planet.

Even this global presence is dwarfed by the varied roles of foam in the universe, from ethereal and conceptual to firm and solid. Foam was present before cosmic birth and when the universe was young; it is evident in the structure of asteroids, supports humankind's efforts to explore the universe, and defines how the galaxies themselves—those mighty multibillion collections of stars, tens of thousands of light-years across—are arranged in space.

seven

COSMIC FOAM

QUANTA, COMETS, AND GALAXIES

An alien space-farer examining our planet through a telescope could discern sea foam on Earth, and might even see how that foam is made. But when we look outward into the universe, our biggest telescopes do not immediately reveal the presence of foam. Nor does foam appear in photographs snapped by NASA spacecraft as they roam the solar system.

Yet there are cosmic foams and bubbles, difficult or impossible to see, that exist in surprising variety and quantity. They are found at the smallest sizes we can imagine, tinier than a subatomic particle; and at the biggest scales we know, the billions of light-years of the entire universe. They are made of galaxies and space, of matter and antimatter, and of curved space and time.

In contrast to Earth where liquid foam is prevalent, liquid foams are the least likely cosmic candidates. Liquids do not exist in space, and there is, to our knowledge, precious little wetness on any celestial bodies, other than our own planet. Water may once have flowed on Mars, but no longer; and although there may be an ocean on Jupiter's satellite Europa, it is hidden under ice and does not heave and froth like an earthly sea. Perhaps there are bubbles within the cold liquid hydrogen found on the planets Jupiter and Saturn, or in oceans yet to be discovered on planets circling distant stars (astronomers have identified over twenty-five such worlds), but that is only speculation.

However, we believe solid foam may exist in the form of pumice on planets and moons within our solar system. There may well be pumice on Venus, thought to be highly volcanic; on Io, the moon of Jupiter where volcanic eruptions have been observed (although the magma on Io seems to be insufficiently viscous to turn into pumice); and on Mars, with its mighty dead volcanoes, such as Olympus Mons, the biggest-known volcano in the solar system. Even if pumice does not exist, vesicular rocks definitely have been found or seen on the Moon and Mars. In an intriguing twist, the meteoritic impacts that produce such rocks sometimes break free pieces of relatively undamaged native rock, at speeds great enough to leave their home planet.

About fifteen such chunks, half a ton's worth, reach the Earth from Mars each year. The most famous is ALH84001, the meteorite found in the Antarctic in 1984, and announced with great fanfare in 1996 as showing evidence for ancient Martian life. That claim has come under heavy fire; but even if it is incorrect, ALH84001 may carry other important information. Geochemist Robert Bodnar, at the Virginia Institute of Technology, has closely examined it and another meteorite from Mars, and found tiny cavities only micrometers across filled with clear liquid. The liquid is evident because it contains a minute gas bubble that can be seen to move under a microscope, like the air bubble in a carpenter's level. Bodnar thinks it is liquid carbon dioxide, and that its presence gives significant information about how specific compounds of carbon and oxygen were formed on Mars, with implications for the possibility of the evolution of life there.

ALH84001 weighs only four pounds, but astronomers have discovered a much bigger piece of material that seems extremely porous. In June 1997, NASA's Near Earth Asteroid Rendezvous (NEAR) spacecraft flew within 760 miles of an asteroid named 253 Mathilde. Asteroids are the millions of rocky chunks orbiting the Sun, mostly between the tracks of Mars and Jupiter. Once it was thought that this swarm is the remnants of an exploded planet, but now it is believed to be pieces of primordial rock that failed to unite into planets in the early solar system. Since these chunks of matter go back to that distant time, scientists want to know all about them.

With an average diameter exceeding thirty miles, Mathilde is a big asteroid, which gave it a measurable gravitational pull on NEAR. Measurements of this force and of Mathilde's size and shape yielded a surprisingly low value for the asteroid's density—much less than that of solid rock, and only slightly greater than that of water. Mathilde's density indicates that it is full of voids—in other words, a solid foam. The likeliest explanation is that Mathilde is made of rocks held loosely together by their mutual gravity; either it was originally formed that way, or repeated collisions have battered it into a pile of floating rubble. Or it may be that the voids are not empty but contain low-density material such as water ice. Perhaps high porosity is characteristic of matter that formed in the early solar system.

CATCHING COSMIC DUST
AND KEEPING ROVER WARM

Spacecraft such as NEAR could not explore the cosmos without the use of man-made versions of foam. These light materials are a godsend to the designers of spacecraft and their payloads, every ounce of which must be painfully hauled off the Earth by sheer brute force. Aerogel is one remarkable example. Peter Tsou of NASA's Jet Propulsion Laboratory (JPL), who makes aerogel out of silicon dioxide, the main component of glass, notes that the material holds world's records in fourteen physical properties. That combination of capabilities has made aerogel essential for two major NASA enterprises. One is Stardust, the first mission planned to return samples from space since Apollo 17 brought back Moon rocks in 1972; it will retrieve bits of a comet, the primordial material that made our solar system, and return them to Earth. The other is Pathfinder, the successful 1997 mission to Mars.

STARDUST

By the time you read this, the Stardust spacecraft, launched by NASA in February 1999, will be flying toward its rendezvous with a comet called WILD-2 (named after the astronomer Paul Wild, who discovered it in 1978, and pronounced VEELD-2). This comet was chosen because it is relatively accessible. When the spacecraft gets within sixty miles of the comet in the year 2004, 240 million miles from Earth, it will deploy panels of aerogel. Small particles from the comet's coma—the cloud of gas and dust surrounding its solid core—will slam into the aerogel at speeds of 14,000 miles per hour. Among all known materials, aerogel is the only one that can bring these tiny hyperbullets to a screeching halt with little damage. Like flies in

A flower insulated from a flame by a piece of aerogel.

amber, the particles will remain safely cradled until the year 2006, when the spacecraft returns to the vicinity of Earth. Then a capsule containing the aerogel will descend by parachute into the Utah desert; the aerogel will be retrieved and its trapped particles removed for study.

Sometimes called "frozen smoke," aerogel is difficult to make, and it has taken considerable time to fabricate pieces of sufficient quality to send on the Stardust mission. That effort has also produced one nearly perfect four-inch cube that brings home the uniqueness of the material.

Imagine smoke of a light, rarefied blue-gray tint. (Thinner pieces of aerogel are so transparent they are practically invisible.) But instead of spiraling lazily upward or following the nearest air current, this particular smoke has been coaxed into a four-inch cube with sharp corners and edges, and persuaded to stay there without spreading out in the least. In contrast to its sheerness, it not only holds its shape but is incredibly strong—a person-sized block of aerogel weighing a pound or two can hold half a ton.

The plan to use aerogel to sample the dust within a comet was named Project Stardust by Donald Brownlee, its principal investigator. Brownlee, an astronomer at the University of Washington in Seattle, has made a career out of space dust. For his senior project at the University of California, Berkeley, he studied cosmic dust floating in the Earth's stratosphere, using a balloon that lofted up some twenty miles. That launched a full-fledged and successful line of research for Brownlee, who has won numerous scientific awards, and has both an asteroid and those atmospheric particles named after him.

The tiny agglomerations of atoms making up the dust particles Brownlee and others study may seem a trivial part of the cosmos. As inhabitants of a semisolid planet, we understandably think the universe revolves around such dense collections of matter. In fact, there is not much bulk material out there, even including the possibility of enormous numbers of undiscovered planets and dead stars that have turned solid. Active stars are made of extremely hot gas. The space between them is filled with cold gas, mainly hydrogen, and with dust—and with dark matter, the recently discovered and invisible celestial material of unknown constitution, thought to fill 90 percent or more of the universe.

Cosmic dust, among other things, reflects the life and death of distant stars. Scientists can tell whether a particular bit of dust comes from a star other than our own Sun. Such particles are ejected by a supernova, which is among the most powerful cosmic events; it is the last insanely violent gasp of a massive dying star, a vast explosion that spreads stellar material over a huge area. The remnants of one such explosion can be seen in the tendrils of the Crab Nebula, a glowing mass of gas and dust still expanding nearly a millennium after the event, observed in 1054 C.E. by Chinese astronomers. The sheer energy of a supernova is spectacular, and supernovas are essential in creating many of the 100-odd chemical elements making up the universe, a role that can be clarified by examining the dust they eject.

Space dust also has meaning closer to home, because our own solar system is believed to have formed from a cloud of dust and gas shaped like a compact disc. Gravity pulled hydrogen gas inward toward the center of the disc. There it became hot and dense enough to flare into the Sun, generating energy through the fusion of hydrogen nuclei into helium. Meanwhile, particles of dust were attracting each other through their mutual gravity, accreting into pebbles, then boulders, then chunks several hundred miles across, and finally full-fledged planets. Not all the planets come wholly from dust. The outer ones like Jupiter are thought to have a central rocky core surrounded by primordial gas, some of which has liquefied. But on our own planet (and on Mercury, Venus, and Mars), most of the material and most of the chemical elements began as bits of dust.

The fundamental science to come out of Stardust arises from the connection between comets and primordial dust. "We'll learn about the nature of the particles that form comets," says Brownlee. "These are the solid materials existing in the outer fringes of the planetary system when the Sun and planets formed." Comets are "dirty snowballs," a picturesque way of saying they are mixtures of ice and dust that came together when the solar system was young. Comets typically occupy long skinny orbits that mostly keep them distant from the Sun, but which bring them back to its vicinity periodically—like Halley's comet, which returns every seventy-six years. So the dust in an orbiting comet remains safely refrigerated until the comet nears the Sun, where the Sun's warmth melts some of the ice to release particles of dust.

To retrieve the dust, there is Project Stardust, which cost $200 million and involves 200 people. This large effort— stretching from Brownlee's office in Seattle to JPL in Pasadena, California; from Lockheed-Martin Astronautics in Denver, which assembled the spacecraft, to the launch site at Cape Canaveral; and continuing for another 3 billion miles out to the comet and back—is devoted to one thing: bringing back pieces of WILD-2. Without aerogel, comet dust could not be harvested, for it's the only substance able to capture dust particles without destroying them. A particle moving twenty or thirty times faster than a rifle bullet carries enough kinetic energy to demolish itself if it smashes into a solid surface; but in aerogel, its energy is dissipated harmlessly. The process is incompletely understood, although it involves the manner in which the speeding particle interacts with strand after strand of glassy silicon dioxide, coming to rest after traversing only a millimeter or two of aerogel.

The exact manner in which the particle is held in aerogel is also miraculous. For reasons still partly obscure, a penetrating particle makes a three-dimensional carrot-shaped track, wide at the entry point, and rapidly narrowing down to a fine point where the particle stops. In other words, the particle might as well be generating a large flashing neon arrow pointing directly to itself. That is an enormous blessing, since the bits of dust are typically 1 to 100 micrometers across and can be seen only under a microscope. Even then, retrieval is possible because of a second blessing: the fact that aerogel is transparent. The material could not be more perfect for Project Stardust.

The expectation is that 1,000 or more comet particles will be harvested in a little over a square foot of aerogel. Like the Moon rocks from earlier space exploration, these particles will be made available to scientists around the world who submit persuasive proposals about what can be learned from them. The pieces of comet will be exposed to every probe scientists can imagine—electron microscopy, spectroscopy with visible and infrared and ultraviolet light, and many more—to examine the fundamental stuff that made our world and the other planets.

Based on what we already know about comets, over 10 percent of what is retrieved will consist of organic compounds, which contain carbon along with other elements. Of all the atomic elements, carbon atoms have the greatest tendency to form complex molecules like DNA and chlorophyll. That's why life on Earth is based on organic compounds, and why we find it hard to imagine other chemical bases for life to arise and thrive. This does not mean that comet dust is expected to harbor life; but it may harbor its precursors. It has been argued that comets brought organic compounds to the early Earth (some scientists even speculate that life itself reached our world by being carried through space). The Stardust samples will surely support some theories about the origins of our world and its life, and destroy others. But the underlying reason for fundamental research like this has been concisely said by Brownlee: "Whatever we find will be interesting."

PATHFINDER

Another cosmic use of aerogel has been found by Gregory Hickey, the JPL engineer charged with designing thermal insulation for the Pathfinder spacecraft that landed on Mars on July 4, 1997, and for the immensely popular Sojourner Rover the spacecraft launched after it landed. This little twenty-three-pound robot investigated the Martian surface, charmingly sniffing one rock at a time—that is, pressing its analytical sensor head against the rock to gather data about its composition.

Insulation was necessary for both vehicles because Mars is cold. Depending on the time of day, the season of the Martian year, and the geographic location, the surface temperature can drop as low as –90 to –100 degrees Celsius (–130 to –150 degrees Fahrenheit). It can also rise up to 70 degrees Celsius higher, but that is still cold. These big temperature swings, as well as the low temperatures themselves, were unhealthy for the electronic innards of Pathfinder and Rover, especially for the batteries backing up the solar cells that powered both vehicles.

Hickey's task was like that of a homeowner who plans on insulating the walls of her house to keep the inside comfortable no matter what the outside temperature. But his insulation had to satisfy stringent constraints, which were especially demanding for the mobile Rover. (Pathfinder remained motionless after landing.) In addition to insulating well, the Rover's insulating material had to have low weight, structural strength (if it required support, that would add weight), and survivability (in case a Martian rock tore a hole in the robot). Weight was a particular obsession of JPL engineers, who expended considerable ingenuity to reduce it in the spacecraft and its payload. The parts that went into the Rover, such as the wheel assembly, were so lightweight, they seemed to float off the hand compared to their conventional counterparts.

Every kind of thermal insulation, including foamed polyethylene, was considered to see how well it would perform on Mars. The different atmosphere there would severely affect how well a foam insulates compared to its performance on Earth. Nothing matched aerogel, which was also pounds lighter than other materials, and so it was chosen to insulate Rover's Warm Electronics Box, which protected everything sensitive to temperature. (Aerogel will also be used in new robot Rovers planned for more elaborate missions to Mars in the next decade.) It would be too much to say that the success of the Sojourner Rover was due solely to aerogel; but without it, along with a host of other clever solutions to difficult problems, Rover would not have been possible.

"IT'S BUBBLES ALL THE WAY DOWN"

Gossamer-light and transparent as aerogel is, there are other natural foams and bubbles, infinitely less tangible and harder to see, that are part of the workings of the cosmos. They are omnipresent in a way reminiscent of a well-known anecdote supposedly rooted in East Indian mythology. Like many another good story, it is probably apocryphal, and has been ascribed to luminaries from Samuel Johnson and William James to Bertrand Russell and Richard Feynman. It goes like this: An anthropologist, seeking cosmological views from a wise tribal elder, is told that the world rests on an elephant, which is supported in turn by a turtle. "All right," says the scientist, "but what supports the turtle?" "Another turtle," he is told. Thinking to make a point, the anthropologist presses on: "But what lies underneath that turtle?" "Oh, my friend," says the elder, "that's easy. It's turtles all the way down."

Today, our own culture's scientific elders might say, "It's bubbles all the way down." Bubbles and foam appear constantly in scientific thinking about the nature and origin of the universe, from its least parts to its mightiest. Later we will discuss how the big bang theory views the cosmos as an expanding bubble of space and time that is now billions of light-years across. In a variant of that theory, our universe may be only one of many, each with its own unique space and time—a foam of universes.

But on the smallest level of cosmic foam, there are tiny regions ruled by the laws of quantum physics, which induce exotic new kinds of foam and bubbles that may have given birth to the entire universe.

The quantum universe is an eerie submicroscopic world where common sense is suspended. It is a place where energy comes not in a smooth flow but in distinct packages termed *quanta;* where solid matter undulates like ocean waves; where particles of light blast straight ahead like missiles. Its bizarreness reaches a peak in the Heisenberg uncertainty principle, which asserts that some physical facts can never be exactly known; and that random variations called *quantum fluctuations* can make matter appear out of nothing. It is a world whose wildness we have tamed and used to some extent, as in solid-state devices, yet whose essential strangeness still lies beyond us. Einstein firmly rejected its inherent randomness when he said, "I shall never believe that God plays dice with the world."

Nevertheless, quantum fluctuations exist and cause some of the weirder actions in the universe. They mean that even in what we call vacuum—the complete absence of anything—electrons can randomly pop into existence, live for a brief instant, and then disappear. The only proviso is that each electron must be accompanied by its antiparticle, a positron, identical but with opposite electrical charge (in this case, positive). (Every kind of elementary particle—electron, proton, and so on—has its "mirror image" antiparticle.) This rapid appearance and disappearance in quantity, reminiscent of the frothy pop and fizz of soda water, is one form of quantum foam.

It is also theorized that just as a storm-tossed sea makes round bubbles of air and water, so quantum turbulence creates curved bubbles of space and time. That happens at a minute scale which originated in the thinking of Max Planck, the physicist who introduced the quantum in 1900. He considered three quantities that define ultimate properties of the universe: the speed of light; his own Planck's constant, which sets the (extremely small) size of quantum effects; and the gravitational constant, which determines the strength of gravity between any two objects. From them, he derived a distance now called the *Planck length*, which seems to be the smallest distance that has physical meaning.

The numerical value of the Planck length is 1.6×10^{-35} meter—so small a dimension that a proton, already unimaginably tiny, looms over it in the same ratio that the acreage of the state of New Jersey looms over a single proton. That striking image comes from the distinguished American theoretical physicist John Wheeler, who first foresaw that what he called *quantum foam* should exist at the Planck scale. In 1962 Wheeler—who worked with Niels Bohr, a founding father of quantum theory, and collaborated with Edward Teller and others at the Los Alamos Scientific Laboratory to develop the hydrogen bomb—came up with the concept of quantum foam, which is based on a combination of quantum mechanics and Einstein's theory of general relativity. To understand what quantum foam is, we must first discuss how general relativity redefined our knowledge of gravity.

Einstein's 1905 theory of special relativity drew the astonishing conclusions that space and time depend on the observer and are linked into a spacetime continuum; that nothing can exceed the speed of light; and that matter and energy can turn into each other. Having established this new view of the universe (later proven correct by experiments and finally by the atomic bomb), Einstein used it to develop general relativity, a new vision of gravity he published in 1915.

Gravity had supposedly been understood ever since Isaac Newton said it is a force between any two objects that becomes stronger as the objects get closer. But although Newton's theory seemingly works well in most situations, it is inconsistent with the rules of special relativity. Einstein proposed instead that gravity is not a force but a matter of geometry. It arises whenever the spacetime continuum—the four-dimensional quilt woven from three spatial dimensions and a temporal one—is curved by the presence of matter.

According to Einstein, a planet may seem to be pulled toward a star such as our Sun by a gravitational force, but what really happens is that the massive Sun dents the surrounding spacetime, like a bowling ball placed on a tightly stretched rubber sheet. The heavy ball makes a depression, and a marble placed on the sheet rolls down the incline toward the ball. Similarly, an enormous Sun deforms spacetime, and a nearby planet rolls down that "hill" and toward the Sun, or, given a sideways push, swings into orbit around it. In both cases the planet moves as if it were attracted by a gravitational force. But according to general relativity, gravity is a direct outcome of the shape of space and time.

Experiments show that general relativity correctly describes gravity, which determines the biggest cosmic structures; and so it is properly called the theory of the large. At the opposite limit, there is quantum mechanics, the theory of the small. Quantum mechanics explains three of the four types of force that define the workings of the universe: the weak force, which governs certain activities in atomic nuclei such as the ejection of electrons; the strong force, which binds protons and neutrons into nuclei; and electromagnetism, which appears in light and radio waves and infrared heat, and which binds atoms and molecules into ordinary matter.

But quantum mechanics does not yet explain gravity, the fourth fundamental force. A theory of quantum gravitation would merge understanding of the large and the small into a full picture of the universe. Although that theory still eludes the best physics minds of our time, Wheeler's study of quantum mechanics and general relativity has led to notable results. One outcome is greater understanding of the spacetime bubble called a *black hole,* a term Wheeler coined in 1967.

A black hole occurs when matter becomes exceedingly dense, as can happen at the heart of a dead star. The powerful gravity that results further compresses the matter into what physicists call a *singularity,* a state of infinitesimal size and infinite density. According to general relativity, that curves the surrounding spacetime so severely that it forms a separate bubble pinched off from ordinary spacetime. Events within that isolated region, such as the emission of light, are invisible to the rest of the universe, a feature captured by the phrase *black hole.* Nevertheless, a black hole can be detected through the gravity it exerts and the matter it pulls in, and astronomers believe they have found several.

THE FOAMY SEAS OF PLANCK

Another result of Wheeler's investigation of quantum gravitation is his idea of quantum foam, which, if he is right, underlies all reality. Quantum foam unites three bedrock ideas: Einstein's insight that powerful gravity and highly curved spacetime go together; the inherent randomness of quantum fluctuations; and the fact that gravity becomes strong enough to affect the quantum world only at the extremely small distances of the Planck length.

Combine these elements, and you get Wheeler's image of spacetime at the smallest scale. As Wheeler explains it, in the normal world, spacetime is "glassy smooth"; but at the Planck scale the world is a "vision of turbulence" full of quantum foam, which is "spacetime itself churned into a lather of distorted geometry. . . . there would literally be no left and right, no before and no after. Ordinary ideas of length would disappear. Ordinary ideas of time would evaporate."

The idea of spacetime frothing and bubbling at the limits of comprehension was also developed in the 1980s by English physicist Stephen Hawking. In an article titled "Quantum Gravitational Bubbles," Hawking offered this concise description: "Spacetime is expected to have a "foamlike" structure on scales of the Planck length or less with high curvature. . . . This foam can be thought of as being built out of . . . gravitational bubbles."

One way to picture quantum foam is to represent the spacetime continuum by the surface of the sea as it is whipped into froth by a powerful hurricane. Look at that frantic activity from a great distance, and you miss the turmoil. Except that the sea is white, it seems undisturbed. But come closer and look through a powerful lens; now you see the individual bubbles that agitate the surface, popping unpredictably into and out of existence at a great rate. Similarly, spacetime looks serene and unchanging from the heights of ordinary existence. Focus down to the minute scale of the Planck length, however, and you see random bubbles of spacetime form, change, and collapse. (Since all normal ideas of space—and time—are utterly lost at that scale, it is meaningless even to imagine any distance shorter than the Planck length.)

There is one essential difference between the oceanic image and quantum foam itself: There is no quantum hurricane—that is, there is nothing analogous to a storm that agitates the ocean. Quantum foam arises solely from the fluctuations at the heart of the quantum world. They are too weak and short-lived to affect ordinary life, but at extremely small scales, they combine with gravity to stir up the fundamental shape of the world, and may have more than tiny effects. It may be that the universe began in a quantum bubble, the starting point of the big bang, the accepted scientific description of how the cosmos has developed since its birth some 15 billion years ago.

BLOWING UP THE BIG BANG

The big bang is visualized as an outward eruption of reality that created all space and time, matter and energy. The fact that the universe is still expanding (first determined by the astronomer Edwin Hubble in 1929) is one indication that the bang happened. A second marker is the cosmic background radiation (CBR), a particular set of electromagnetic waves that fills the sky. Invisible to the eye, it can be detected by antennas operating at the short radio wavelengths called *microwaves*. (CBR was accidentally discovered in 1965 by Nobel laureates Arno Penzias and Robert Wilson of Bell Laboratories, as they tested radio communications with artificial Earth satellites.) These electromagnetic waves have exactly the right wavelengths to be left over from the big bang, taking into account how they would have been stretched by the expanding universe.

There is other evidence supporting the big bang, but even so, scientists are wary about such an all-encompassing theory. Like canny used-car buyers who kick the tires and check for rust, they probe for weak spots by asking questions and seeing how well the theory answers them. Sure enough, there are issues the big bang theory does not address as well as it might. It simply assumes the universe began and expanded, never quite going back to the very first instant to explain how it all originated. And there is the "horizon problem," posed by the fact that the CBR is the same no matter where it originates in the sky. It doesn't matter if your antenna points straight up from the North Pole or straight down from the South Pole, toward the constellation Orion or directly opposite; it registers exactly the same incoming wavelengths and strength to an accuracy of 1 part in 100,000.

Electromagnetic radiation comes from hot bodies, and so if the CBR is the same everywhere, the temperature of the universe must also have been uniform in that ancient era when light first filled in. For a body to have a uniform temperature, heat has to spread evenly through it. When the CBR first filled the universe, however, some 300,000 years after the bang, the cosmos was already so big that parts of it were "below the horizon" relative to other parts. That is, they were so far apart that even light at its vast speed, let alone heat, could not have covered the distance between them; therefore, that young universe could not have reached a constant temperature.

This difficulty and others led to the modification of the big bang called *inflationary theory*. Its basic idea came in 1979 to Alan Guth, a young elementary particle physicist. In contrast to the initial version of the big bang, where the universe expanded at a steady rate, Guth's notion was that almost immediately after the universe was born, when it was still minute, it underwent a brief period of extremely rapid expansion and then reverted to a lazier pace. This turned out to be a breakthrough idea that put the big bang on a much firmer basis.

As an example of how inflation resolved difficulties, it eliminated the horizon problem. According to Guth's calculations, just before the rapid expansion the infant cosmos was smaller than a proton. In this tiny bubble, every part was within easy reach of every other part, and there was no difficulty in reaching a common temperature. Once that uniformity was established, it would be maintained during further expansion of the universe, until now we see it in the CBR.

Inflation needed a mechanism to make it happen, and Guth provided that, too. As shown by the quantum fluctuations that produce electrons and positrons, a vacuum is more than absolute nothingness. It may also contain energy, a state called a *false vacuum*. Guth realized that this peculiar condition could have briefly existed in the early cosmos, where it provided a property equally odd but just right to propel expansion—a gravitation opposite to what we know today, repelling matter instead of attracting it. This happened in the merest sliver of time, a hundred-thousandth of a nano-nano-nanosecond; then inflation ended, the universe reverted to a more stately rate of growth, and eventually reached its present size.

Inflation is another way bubbles enter cosmic history. Guth's calculations showed that as the inflationary period came to an end, bubbles of normal matter would develop within the false vacuum, like bubbles of vapor forming within a pot of water just coming to the boil. Then in 1981, the Soviet physicist Andrei Linde showed that the universe we see must lie within one of these bubbles, now enormously grown to a size of billions of light-years. Linde's theory resolved some lingering questions about the inflationary big bang, leaving it as the generally accepted theory of how the universe developed.

The inflationary theory supports an astonishing scenario for the origin of the universe: that it began as a random quantum fluctuation. The idea was put forth in 1973 by physicist Edward Tryon of the City University of New York. (One anecdote has Tryon first blurting out the thought, much to his embarrassment, during a seminar he attended in 1969.) If you need a moment of creation when something arises out of nothing, a quantum fluctuation fills the bill. Under most conditions, quantum fluctuations are too short-lived to form the basis for a whole cosmos, but Tryon pointed out a way that a fluctuation could survive for much longer. That still left a problem, however: The newborn universe would be so small, with such strong gravity, that it would soon collapse on itself anyway. Guth's idea of inflation provides a way for this mustard seed of reality to fluff itself outward into a full cosmos, turning randomness into reality.

The possibility of this random origin of the universe is now taken extremely seriously by physicists and cosmologists. If it truly is "bubbles all the way down," inflation is what links them, expanding an infinitesimally small bubble of spacetime into the enormous bubble we occupy, the universe (or as some theorists have proposed, a multiverse, many universes clustered together like bubbles in a foam, each with its own particular spacetime and physical reality).

Depending on your temperament and beliefs, the thought of a random fleck of quantum foam turning into the mad variety of everything we know, atoms and galaxies, rocks and people, cabbages and kings, is frightening or amusing, offensive or liberating. (Tryon has been quoted as saying, "The Universe is simply one of those things that happens from time to time.") Whatever your reaction, these large and small bubbles of space, time, and reality are compelling images, and the theories they represent are our best scientific attempts to explain the universe.

But we should not forget that these theories, like any theories, are intellectual constructs; and that the bubbles they describe are too small, or too vast, or were formed too long ago to be observed directly. Moreover, the theories predict little that can be tested by observation or experiment, which makes it difficult to check them against reality. Finally, there is always the possibility that as we look more deeply into the universe, we may find the hard fact that upsets the most persuasive of theories.

In fact, what has been called the scientific breakthrough of the year 1998 has cast some doubt on current ideas. The breakthrough came when astronomers observing distant exploding supernova stars found them to be dimmer than expected. This indicated that the stars had been carried a surprisingly great distance by the expansion of the universe, meaning that the outward thrust is speeding up. But in the scenario of the big bang with inflation, the gravity due to all the matter that exists tends to pull the cosmos inward, slowing down or even reversing its growth, not accelerating it. The contradiction suggests that the universe is permeated with a kind of antigravity force that pushes matter apart. By suggesting the existence of this mysterious component, the new observations put present theories of the universe into question and remind us that some cosmic bubbles may resemble soapsuds more than we knew—that is, they are fragile and easily popped.

MAPPING THE BUBBLES

There is, however, one foam of truly cosmic proportions that is also truly observable. It consists of enormous bubbles whose skins are made of galaxies linked together by gravity, first discerned in 1986 by the astronomers Margaret Geller and John Huchra of the Harvard-Smithsonian Center for Astrophysics. They had to record and analyze thousands of images of distant galaxies before the foamy pattern became clear. Although telescopic observations show where a galaxy lies in the sky, they do not immediately reveal its distance from the Earth. It takes more to determine the third spatial dimension, and to turn that knowledge into images of curved bubbles.

Ancient astronomers did not know about galaxies and could not measure how far celestial objects lay from us, although they located them in the sky through the astronomical equivalents of latitude and longitude. But gradually we developed methods of measuring the distance from Earth to the Sun and planets of the solar system; to other stars in our galaxy (up to thousands of light-years away); to other galaxies (up to hundreds of millions of light-years away); and finally to the farthest objects we know, the mysterious and powerful quasars (up to 15 billion light-years away).

Not surprisingly, the greatest distances were the last to be

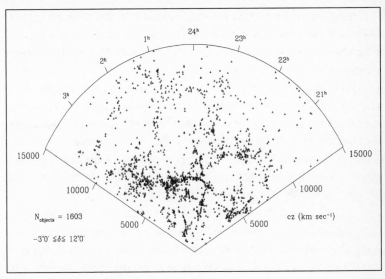

A slice through the universe, stretching 500 million light-years from Earth at the tip of the wedge. Each dot is a galaxy.

measured. When Edwin Hubble established that final distance scale in the 1920s, he carried us into a new universe full of galaxies like our own Milky Way. These had been observed since the seventeenth-century as dim ill-defined areas of light, but no one knew what they were; they were simply called *nebulas,* which means "cloud." Well into this century, some astronomers believed they were relatively small and lay within our own galaxy, which constituted the entire universe; others thought they were huge "island universes" much farther away.

Using the new Mount Wilson telescope near Pasadena, California, then the world's biggest with its 100-inch mirror, Hubble settled the issue in 1924 when he announced that the Andromeda nebula lay 1 million light-years away (later corrected to over 2 million). That turned Andromeda and most other nebulas into galaxies (some nebulas indeed lie within our own Milky Way)—each containing billions of stars and stretching over thousands of light-years—because only such stupendous objects could be seen at such distances; it also told us that we live within a universe much bigger than anyone had imagined.

Hubble went on to find that this enormous universe, dappled with what is estimated to be 100 billion galaxies, is relentlessly growing. It had been known for some time that the light from each galaxy is "redshifted"; that is, its wavelengths are stretched toward the red, or long-wavelength, end of the color spectrum. From the general properties of light, we know that this means that each galaxy is moving away from us. As Hubble's further analysis showed, it also means that each galaxy is moving away from every other one in an expanding universe.

Hubble measured redshifts for many galaxies, and from that he determined how fast each one is fleeing from us. Then in 1929, he made his key discovery, Hubble's law, which states that the speed of each galaxy is proportional to its distance from our own galaxy. That is exactly what you would expect if you modeled an expanding universe by painting polka dots on the surface of a balloon and then blowing up the balloon. If an intelligent bug with astronomical pretensions were standing on each polka dot, each bug would see every other dot moving away from its own. Further, every remote dot would move at a speed proportional to its distance from each observing bug, exactly as an observer in each galaxy sees every other galaxy fleeing from his own, carried along by the expansion of the cosmos.

Hubble's expanding universe formed the basis of the big bang idea, and his law gave a new probe of the deep cosmos, since it tells us how to convert a measured degree of redshift into a known distance. Among the half-dozen different tape measures astronomers use, Hubble's stretches the longest, up to billions of light-years. That's the final piece of information needed to specify exactly where a given galaxy lies in space, and so to address this fundamental question: How are these immense concentrations of matter spread throughout the universe?

The cosmic background radiation suggests one answer. Its homogeneity indicates not only a uniform temperature in the young universe—the issue resolved by Guth's inflationary idea—but a uniform everything, including matter. If the universe began that way, it would evolve similarly, producing galaxies homogeneously sprinkled throughout space without any special pattern. Although this conclusion seems reasonable, it needs to be tested. After all, it once seemed just as reasonable to think the Earth was flat. Only extensive exploration and mapmaking uncovered its true shape.

The comparable cosmic exploration means determining the three-dimensional distribution of galaxies by measuring the size of their redshifts and hence their distances from the Earth. In Hubble's day, many hours were needed to find the redshift for a single galaxy. That was the time required for the weak light from the galaxy to trickle through the telescope and form a usable image on a photographic plate. In the 1980s, when John Huchra and Margaret Geller began measuring redshifts, they used sensitive electronic detectors of light that reduced the time to minutes per galaxy.

The results did not immediately appear in three dimensions. They came in the form of thousands of dots, each representing a galaxy. These dots were sprinkled over flat images representing slices cut through the universe at different angles, each extending 500 million light-years from the Earth. What the eye sees immediately is that the dots are not strewn evenly but are concentrated more or less in the form of adjoining circles of different sizes. Each circle of galaxies surrounds an empty space up to 200 million light-years across with hardly a galaxy in it.

Geller turned these two-dimensional plots into a coherent three-dimensional cosmic picture by comparing the arrangement of galaxies in space to a mass of soap bubbles in a kitchen sink. Each soap bubble consists of a film made of soap molecules and enclosing a volume of air; each cosmic bubble consists of a film made of galaxies and enclosing a volume of space. The soap molecules are held in place by chemical forces, which make them act as surfactants; the galaxies are held in place by their mutual gravitational attraction. Geller's model is a great help in understanding the three-dimensional cosmic foam, which is far too big to see in its entirety through a telescope.

Her insight came from knowing that a cross section through a bubble is a circle; that is, if you could carefully slide a sheet of paper through a soap bubble, you would see a circle where the two intersected. Now slide your paper through a mass of soapsuds, and as it cuts through adjoining bubbles of varying sizes, it generates adjoining circles, also of varying sizes. Tilt the paper so it intersects the foam at a different angle, and you get a different pattern of circles. This is probably not too difficult to see in your mind's eye; Geller, however, carried out the harder task of imagining the three-dimensional shape from her set of two-dimensional images.

The result showed that galaxies are found almost entirely on the surfaces of immense spheres hundreds of millions of light-years across—not the even dispersal suggested by the unvarying CBR. Still, we know in principle how this distribution might have formed. Gravity could have created it, but only if matter were spread throughout the early cosmos like lumpy oatmeal rather than smooth gruel. Each denser lump would generate stronger gravity; and so, like the rich getting richer, the clots would grow as they attracted matter while less dense regions would diminish, until all matter had sculpted itself into a pattern such as the bubbles we see. The smoothness of the CBR seems to deny that essential initial lumpiness, but this may be only a matter of degree, since the CBR may vary by up to 1 part in 100,000. The role of dark matter, which exerts gravity although it is invisible, must also be considered.

While cosmologists and astrophysicists ponder these issues, redshift measurements continue. The 15,000 galaxies Geller and Huchra examined are only a tiny fraction of the universe, and it is essential to see if their pattern extends farther. The more distant the galaxies we scrutinize, the longer the time their light has traveled and the more closely we approach the era of the big bang, so that, says Geller, "we can really observe the history of the universe directly." Geller and her colleagues plan to examine 50,000 additional galaxies, and the Sloan Digital Sky Survey project, sponsored by the Alfred P. Sloan Foundation and other sources has installed a telescope in New Mexico to find the distances to 1 million galaxies. These extended measurements may establish such characteristics as whether the cosmic foam is closed-cell or open-cell—that is, truly like soapsuds with separate bubbles, or like a sponge with interconnected voids. The distinction may seem minor compared to the sheer size of the foam, but may be just the feature that distinguishes among different theories to help us determine the shape, structure, and history of the universe.

THE MARVELOUS DIVERSITY OF FOAM

The universe may have begun from a quantum fluctuation, and when it was very young, consisted of bubbles of normal matter embedded in the false vacuum. Fifteen billion years later, we find ourselves within a stupendous galactic foam, while around us a minute quantum foam imperceptibly shapes spacetime. This span in the meaning of foam is awesome to the spirit but distant to the senses, for it represents extremes of space and time that humans cannot grasp.

If foam marks the exotic limits of the cosmos, it also represents the miraculous variety of the world around us. We respond best to foam when it appears within reach of our senses.

For me, that response came not long ago when I drove north from San Francisco to the Point Reyes lighthouse, perched precariously on the rocks. Clambering around the site, I saw and heard foam being made as waves smashed into the rocks, flinging spray into the air and beating the water into white turbulence. Farther out, where the water was calmer, it displayed long-lived streamers of a different foam, tinted off-white by tiny seashells. Beyond that, foam appeared again as lines of whitecaps on the open sea.

Looking farther north along the deserted sweep of Point Reyes beach, I saw foam on the grand scale. It appeared as a miles-long scalloped line of white glowing against the dark sand, made of the arcs of waves breaking side by side. At that distance, I could not see the chaotic details of the surf, but I knew that it left behind ephemeral traces, bubbles that burst one by one—the fragile aftermath of powerful breakers, bubbles which would play around the feet of anyone who stood near the sea.

And so, although we exist within a galactic foam, our lives are rooted in a world where we hear the foamy roar of the sea and feel its bubbles, where the practical and the aesthetic complement each other. We drink beer and its froth; watch soufflés rise and fall; use foamed plastic peanuts in their trillions; and diagnose disease with bubbles while our bodies follow their own internal cellular patterns. Amid these intimate and approachable levels, we send foamy aerogel deep into space, trying—as we always must—to bridge the gap between the ordinary and the exotic, the human and the cosmic, between what we know and what we want to know.

bibliography

Abel, Bob. *The Book of Beer*. Chicago: Regnery, 1976.

Almgren, Frederick, and Jean Taylor. "The Geometry of Soap Films and Soap Bubbles." *Scientific American*, July 1976, 82–93.

Andrés-Lacueva, Cristina, et al. "Characteristics of Sparkling Base Wines Affecting Foam Behavior." *Journal of Agricultural and Food Chemistry* 44 (1996): 989–95.

Aubert, James H., Andrew M. Kraynik, and Peter B. Rand. "Aqueous Foams." *Scientific American*, May 1986, 72–82.

Barber, Bradley P., Robert A. Hiller, Ritva Löfstedt, Seth J. Putterman, and Keith R. Weninger. "Defining the Unknowns of Sonoluminescence." *Physics Reports* 281 (1997): 65–143.

Bartusiak, Marcia. "Mapping the Universe." *Discover*, August 1996, 60–63.

Bergquist, Patricia. *Sponges*. Berkeley: University of California Press, 1978.

Bernard, A., D. Demaiffe, N. Mattielli, and R. S. Punongbayan. "Anhydrite-Bearing Pumices from Mount Pinatubo: Further Evidence of Sulphur-Rich Silicic Magmas." *Nature*, November 14, 1991, 139–40.

Bernatowicz, Thomas J., and Robert M. Walker. "Ancient Stardust in the Laboratory." *Physics Today*, December 1997, 26–32.

Bikerman, J. J. *Foams*. New York: Springer-Verlag, 1973.

Boden, Margaret A., ed. *Artificial Intelligence*. San Diego: Academic Press, 1996.

Borzeszkowski, Horst-Heino von, and H.-J. Treder. *The Meaning of Quantum Gravity*. Dordrecht, Holland: Reidel, 1988.

Boys, C. V. *Soap Bubbles, Their Colours and Forces Which Mould Them*. 1890. Reprint, New York: Dover Publications, 1959.

Brillat-Savarin. *The Physiology of Taste, or, Meditations on Transcendental Gastronomy*. 1825. Reprint, New York: Dover Publications, 1960.

Brissonnet, F., and A. Maujean. "Identification of Some Foam-Active Compounds in Champagne Base Wines." *American Journal of Enology and Viticulture* 42 (1991): 97–102.

Brown, Malcolm W. "Chemists Create Foam to Fight Nerve Gases." *New York Times*, March 16, 1999.

Cornell, James, ed. *Bubbles, Voids, and Bumps in Time: The New Cosmology*. New York: Cambridge University Press, 1989.

Crum, Lawrence A. "Sonoluminescence." *Physics Today*, September 1994, 22–29.

Davis, Richard A. *Principles of Oceanography*. Reading, Mass.: Addison Wesley, 1977.

Dean, Cornelia. "Mapping the Beach, One Grain at a Time." *New York Times*, October 21, 1997.

Durian, D. J. "Bubble-Scale Model of Foam Mechanics: Melting, Nonlinear Behavior, and Avalanches, " *Physical Review E* 55 (1997): 1739–51.

Durian, D. J., D. A. Weitz, and D. J. Pine. "Multiple Light-Scattering Probes of Foam Structure and Dynamics." *Science*, May 3, 1991, 686–88.

Elias, Florence, Cyrille Flament, Jean-Claude Bacri, Olivier Cardoso, and François Graner. "Two-dimensional Magnetic Liquid Froth: Coarsening and Topological Correlations." *Physical Review E* 56 (1997): 3310–18.

Elias, Hans-Georg. *Mega Molecules: Tales of Adhesives, Bread, Diamonds, Eggs, Fibers, Foams, Gelatin, Leather, Meat, Plastics, Resists, Rubber, and Cabbages, and Kings*. New York: Springer-Verlag, 1987.

Fenichell, Stephen. *Plastic: The Making of a Synthetic Century*. New York: HarperBusiness, 1996.

Finegan, Jay. "Down in the Dump." *Inc.*, September 1990, 90–100.

Fisher, Richard Virgil, Grant Heiken, and Jeffrey B. Hulen. *Volcanoes: Crucibles of Change*. Princeton: Princeton University Press, 1997.

Fricke, Jochen. "Aerogels." *Scientific American*, May 1988, 92–97.

Friendly, Alfred. *Beaufort of the Admiralty: The Life of Sir Francis Beaufort*. New York: Random House, 1977.

Geller, Margaret, and John Huchra. "Mapping the Universe." *Science*, November 17, 1989, 897–903.

Gibson, Lorna J., and Michael Ashby. *Cellular Solids: Structure and Properties*. New York: Cambridge University Press, 1999.

Glanz, James. "Exploding Stars Point to a Universal Repulsive Force." *Science*, January 30, 1998, 651.

Gonatas, C. P., J. S. Leigh, A. G. Yodh, James A Glazier, and B. Prause, "Magnetic Resonance Imaging of Coarsening Inside a Foam." *Physical Review Letters* 75 (1995): 573–76.

Gopal, A. D., and D. J. Durian. "Nonlinear Bubble Dynamics in a Slowly Driven Foam." *Physical Review Letters* 75 (1995): 2610–13.

Guth, Alan. *The Inflationary Universe*. Reading, Mass.: Addison Wesley, 1997.

Hawking, S. N., D. N. Page, and C. N. Pope. "Quantum Gravitational Bubbles." *Nuclear Physics* B170 (1980): 283–306.

Hoffer, Jerry M. "Identifying Acid Wash, Stone Wash Pumice." *Textile Chemist and Colorist*, February 1993, 13–15.

Hooke, Robert. *Micrographia*. 1665. Reprint, New York: Hafner, 1961.

Howle, Laurens, David G. Schaeffer, Michael Shearer, and Pei Zhong. "Lithotripsy: The Treatment of Kidney Stones with Shock Waves." *SIAM Review* 40 (1998): 356–71.

Illy, Andrea, and Rinantonio Viani, eds. *Espresso Coffee: The Chemistry of Quality*. San Diego: Academic Press, 1995.

Isenberg, Cyril. *The Science of Soap Films and Soap Bubbles*. New York: Dover Publications, 1992.

Janssen, Phillip. *Espresso Quick Reference Guide*. Seattle: Coffee Time Publications, 1995.

Johnson, Julie. "Wrong Sort of Glass Robs Beer of Its Flavour." *New Scientist*, June 17, 1995, 11.

Judge, Michael. "Foam Sweet Foam." *New Scientist*, September 27, 1997, 34–37.

Kanigel, Robert. "Bubble, Bubble." *The Sciences*, May/June 1993, 32–38.

Kaplan, Steven Laurence. "The Times of Bread in Eighteenth-Century Paris." *Food and Foodways* 6 (1996): 283–305.

———. "Breadways." *Food and Foodways* 7 (1997): 1–44.

Katz, Solomon H., and Fritz Maytag. "Brewing an Ancient Beer." *Archaeology*, July/August 1991, 24–33.

Kelvin, William Thompson, and D. L. Weaire, (ed.) *The Kelvin Problem: Foam Structures of Minimal Surface Area*. London: Taylor and Francis, 1996.

Koepke, Peter. "Effective Reflectance of Oceanic Whitecaps." *Applied Optics* 23 (1984): 1816–24.

Koplos, Janet. "Brian Tolle at Basilico Fine Arts." *Art in America*, November 1996, 110.

Kraus, E. B. *Atmosphere-Ocean Interaction*. New York: Oxford University Press, 1972.

Kurti, Nicolas, and Hervé This-Benckhard. "Chemistry and Physics in the Kitchen." *Scientific American*, April 1994, 66–71.

———. "The Amateur Scientist: The Kitchen as Lab." *Scientific American*, April 1994, 120–23.

Leonardo da Vinci and Carlo Pedretti. *The Leonardo da Vinci Codex Hammer*. New York: Christie, Manson and Woods International, 1994.

Liss, Peter S., and Robert A. Duce. *The Sea Surface and Global Change*. New York: Cambridge University Press, 1997.

Lovett, David. *Demonstrating Science with Soap Films*. Bristol: Institute of Physics Publishing, 1994.

McGee, Harold. *On Food and Cooking: The Science and Lore of the Kitchen*. New York: Fireside/Simon and Schuster, 1997.

McKenna, James T. "Woodpecker Damages Shuttle Insulation." *Aviation Week and Space Technology*, June 5, 1995, 66–67.

Medwin, Herman, Jeffrey A. Nystuen, Peter W. Jacobus, Leo H. Ostwald, and David E. Snyder. "The Anatomy of Underwater Rain Noise." *Journal of the Acoustical Society of America* 92 (1992): 1613–23.

Meikle, Jeffrey L. *American Plastic: A Cultural History*. New Brunswick, N.J.: Rutgers University Press, 1995.

Melosh, H. Jay. "Blast Off." *The Sciences*, July/August 1998, 40–46.

Metcalf, Harold. "That Flashing Sound." *Science*, February 27, 1998, 1322–23.

Meyer, Richard E. *Waves on Beaches and Resulting Sediment Transport*. New York: Academic Press, 1972.

Michel, Rudolph H., Patrick E. McGovern, and Virginia R. Badler. "Chemical Evidence for Ancient Beer." *Nature*, November 5, 1992, 24.

Monahan, Edward, and Iognaid G. O'Muircheartaigh. "Whitecaps and Passive Remote Sensing of the Ocean Surface." *International Journal of Remote Sensing* 7 (1986): 627–42.

Moss, William C., Douglas B. Clarke, John W. White, and David A. Young. "Sonoluminescence and the Prospects for Table-Top Micro-Thermonuclear Fusion." *Physics Letters A*, 211 (1996): 69–74.

Newhall, Christopher, and Raymundo S. Punongbayan. *Fire and Mud*. Seattle: University of Washington Press, 1997.

Nunes, Fernando M., Manuel A. Coimbra, Armando C. Duarte, and Ivonne Delgadillo. "Foamability, Foam Stability, and Chemical Composition of Espresso Coffee as Affected by the Degree of Roast." *Journal of Agricultural and Food Chemistry* 45 (1997): 3328–43.

Ouelette, Jennifer. "New Ultrasound Therapies Emerge." *Industrial Physicist*, September 1998, 30–33.

Peterson, I. "Noise at Sea: Cries of Infant Microbubbles." *Science News*, December 1, 1990, 341.

Phelan, Robert, Denis Weaire, and Kenneth Brakke. "Computation of Equilibrium Foam Structures Using the Surface Evolver." *Experimental Mathematics* 4 (1995): 181–92.

Pickard, George L., and William J. Emery. *Descriptive Physical Oceanography*. Oxford: Pergamon Press, 1982.

Pliny the Elder. *Natural History: A Selection*. New York: Penguin, 1991.

Pope, David. "NASA Puts the Heat on Aerogels." *Industrial Physicist*, September 1997, 13.

Powell, Corey S. "War Without Death." *Discover*, April 1999, 29.

Prause, Burkhard, James A. Glazier, Samuel P. Gravina, and Carlo D. Montemagno. "Three-dimensional Magnetic Resonance Imaging of a Liquid Foam." *Journal of Physics: Condensed Matter* 7 (1995): L511–16.

Preston, Kendall, and Michael J. B. Duff. *Modern Cellular Automata*. New York: Plenum Press, 1984.

Putterman, Seth J. "Sonoluminescence: Sound into Light." *Scientific American*, February 1995, 46–51.

Rees, Martin J. *Before the Beginning: Our Universe and Others*. Reading, Mass.: Perseus, 1997.

Rhodes, Richard. *Deadly Feasts: Tracking the Secrets of a Terrifying New Plague*. New York: Simon and Schuster, 1997.

Saïb, Ali, and Hugues de Thé. "Molecular Biology of the Human Foamy Virus." *Journal of Acquired Immune Deficiency Syndromes and Human Retrovirology* 13, suppl. 1 (1996): S254–S260.

Satchell, Michael. "Fight for Pigeon River." *U. S. News and World Report*, December 4, 1989, 27–32.

Scarth, Alwyn. *Volcanoes: An Introduction*. College Station: Texas A&M University Press, 1994.

Schneider, David. "A Blue Note." *Scientific American*, August 1997, 18.

Seife, Charles. "Bubbles Will Bust up Pollutants." *New Scientist*, October 18, 1997, 19.

Seiwert, C. M., and E. Adkins-Regan, "The Foam Production System of the Male Japanese Quail." *Brain, Behavior and Evolution* 52 (1998): 61–80.

Sigmund, Karl. *Games of Life: Explorations in Ecology, Evolution, and Behavior*. New York: Oxford University Press, 1993.

Slocum, Joshua. *Sailing Alone Around the World*. New York: Century, 1900.

Stewart, Ian. "Double Bubble, Toil, and Trouble." *Scientific American*, January 1998, 104–7.

Thompson, D'Arcy Wentworth. *On Growth and Form*. 1917. Reprint, New York: Dover Publications, 1992.

Travis, John. "Helping Premature Lungs Breathe Easier." *Science*, July 23, 1993, 426.

Tryon, E. P. "Is the Universe a Vacuum Fluctuation?" *Nature*, Dec. 14, 1973, 396–97.

Valenti, Michael. "Lawbreakers Beware the Web of Justice." *Mechanical Engineering*, April 1994, 86–87.

Veverka, J., et al. "NEAR's Flyby of 253 Mathilde: Images of a C Asteroid." *Science*, December 19, 1997, 2109–14.

Vogel, Steven. *Life in Moving Fluids: The Physical Biology of Flow*. Boston: Willard Grant, 1981.

Vos, Louis de, et al. *Atlas of Sponge Morphology*. Washington, D.C.: Smithsonian Institution Press, 1991.

Wald, Matthew L. "New Cement System Reins in Runaway Plane." *New York Times*, May 13, 1999.

Wallace, Paul J., and Terence M. Gerlach. "Magmatic Vapor Source for Sulfur Dioxide Released During Volcanic Eruptions: Evidence from Mount Pinatubo." *Science*, July 22, 1994, 497–99.

Weaire, Denis. "Froths, Foams, and Heady Geometry." *New Scientist*, May 21, 1994, 34–37.

Weiss, Robin A. "Foamy Viruses Bubble On." *Nature*, March 21, 1996, 201.

Wheeler, John Archibald. *Geometrodynamics*. New York: Academic Press, 1962.

———. *Geons, Black Holes, and Quantum Foam*. New York: Norton, 1998.

Williams, Howel, and Alexander R. McBirney. *Volcanology*. San Francisco: Freeman, Cooper, 1979.

Williams, Nigel. "How the Ancient Egyptians Brewed Beer." *Science*, July 26, 1996, 432.

Wilson, A. J., ed. *Foams: Physics, Chemistry, and Structure*. New York: Springer-Verlag, 1989.

Wolfram, Stephen. *Cellular Automata and Complexity*. Reading, Mass.: Addison Wesley, 1994.

index

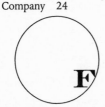

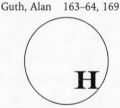

M